Quantum Sensors

Quantum Sensors

Mason Perry

Kruger Brentt
Publishers
2026

Kruger Brentt Publishers UK. LTD.
Company Number 9728962

Regd. Office: 68 St Margarets Road, Edgware, Middlesex HA8 9UU

ISBN: 978-1-78715-625-8

For information on all our publications visit our website at http://krugerbrentt.com/

Preface

Quantum sensing represents one of the most groundbreaking developments in modern science and technology, offering unprecedented precision in the measurement of physical quantities such as time, magnetic and electric fields, temperature, and gravitational forces. By exploiting quantum phenomena—including superposition, entanglement, and tunneling—quantum sensors promise to revolutionize applications in healthcare, navigation, defense, environmental monitoring, and fundamental physics research.

Quantum Sensors provides a comprehensive introduction to the principles, technologies, and practical applications of quantum sensing. The book covers foundational concepts in quantum mechanics relevant to sensing, explores the design and operation of quantum devices, and examines recent advances in experimental and applied quantum technologies. By bridging theoretical insights with real-world applications, the text equips readers with the understanding needed to harness the potential of quantum sensors across multiple domains.

The book is intended for students, researchers, and professionals in physics, engineering, and related fields who seek a thorough understanding of quantum sensing principles and their implementation. By combining theory, case studies, and emerging research, *Quantum Sensors* offers a valuable resource for anyone interested in the forefront of measurement science.

Mason Perry

Contents

CHAPTER 1

Introduction to Quantum Sensors

1.1 WHAT IS A QUANTUM SENSOR?

Quantum sensors represent one of the most cutting-edge advancements in modern science and technology. At their core, quantum sensors exploit the principles of quantum mechanics to achieve sensitivity and precision far beyond what classical sensors can offer. These devices are engineered to detect and measure physical quantities such as time, magnetic fields, gravitational forces, acceleration, and temperature with exceptional accuracy. Their development has ushered in new possibilities across diverse sectors, including medical diagnostics, navigation, geophysics, and fundamental physics research.

1.1.1 Understanding the Fundamentals of Quantum Mechanics

To fully appreciate the working principles of a quantum sensor, it is essential to grasp some basic concepts from quantum mechanics. Quantum mechanics is the branch of physics that deals with phenomena at very small scales, typically at the level of atoms and subatomic particles. Unlike classical mechanics, which assumes continuous and predictable behavior, quantum mechanics is probabilistic and involves discrete energy levels, wave-particle duality, and quantum superposition and entanglement. These properties allow quantum systems to behave in ways that classical systems cannot, and quantum sensors utilize these properties to detect subtle changes in their environment. One of the most important quantum phenomena used in quantum sensing is superposition, which allows a quantum system to exist in multiple states at once. Another is entanglement, where particles become so deeply linked that the state of one instantly influences the state of another, regardless of distance. These characteristics are harnessed in quantum sensors to achieve incredibly high sensitivity and precision. For instance, in atomic clocks—one of the most successful applications of quantum sensing—the superposition of atomic states is used to measure time with such precision that it only loses one second in millions of years.

1.1.2 Types of Quantum Sensors and Their Applications

Quantum sensors come in various forms depending on the physical property they are designed to measure. One common type is the atomic magnetometer, which measures magnetic fields using the spin of atoms. These devices are widely used in neuroscience for magnetoencephalography (MEG), where they map brain activity by detecting the weak magnetic fields produced by neural activity. Atomic magnetometers are also useful in defense applications for detecting submarines and other hidden metallic objects. Another prominent quantum sensor is the atomic gravimeter, which measures gravitational acceleration by analyzing how cold atoms fall under the influence of gravity. This technique, known as atom interferometry, allows scientists to map the Earth's gravitational field with extreme precision. Such measurements are valuable in fields like geophysics, archaeology, and even oil exploration, where subtle changes in gravity can indicate the presence of underground structures or resources.

Quantum gyroscopes are also notable for their role in navigation. Unlike conventional gyroscopes that rely on mechanical rotation, quantum gyroscopes measure rotation based on the interference patterns of atoms or photons in a controlled quantum state. These devices offer navigation solutions in environments where GPS signals are unavailable, such as in submarines or spacecraft. The promise of GPS-independent navigation is particularly appealing for military and aerospace applications.

Quantum-enhanced imaging systems are another significant application area. These sensors utilize the quantum properties of light to enhance imaging resolution and sensitivity. For example, quantum sensors have been proposed for medical imaging systems that can detect disease markers with greater specificity, or for astronomical telescopes that can observe distant celestial bodies with improved clarity by reducing noise in the collected light signals.

1.1.3 Working Principles behind Quantum Sensing

Quantum sensors rely on precise control and measurement of quantum states. Typically, a quantum system—such as an atom, ion, or photon—is prepared in a known quantum state. The system is then exposed to the external physical quantity to be measured, such as a magnetic or gravitational field. This external influence causes a change in the quantum state, and the sensor reads this change to infer the magnitude of the physical quantity. The accuracy of a quantum sensor is determined by how well the quantum system can be isolated from other environmental influences and how precisely the changes in the system's quantum state can be measured. For this reason, quantum sensors are often operated in highly controlled environments such as vacuum chambers or at cryogenic temperatures to minimize noise and decoherence. Advanced technologies like laser cooling and electromagnetic traps are frequently employed to prepare and maintain the quantum states in the desired conditions.

1.1.4 Challenges and Advancements in Quantum Sensing

Despite their impressive capabilities, quantum sensors face several challenges that limit their widespread deployment. One major issue is decoherence, where interaction with the environment causes a quantum system to lose its quantum properties. Maintaining coherence over longer periods or in practical, field-deployable conditions is a major area of research in quantum technology. Additionally, scalability and cost are significant concerns. Many quantum sensors rely on complex laboratory setups, making them less suitable for commercial or mobile applications. However, ongoing advancements in miniaturization and quantum control technologies are gradually overcoming these barriers. Recent developments in quantum materials, such as diamond nitrogen-vacancy (NV) centers, have shown promise in creating more robust and practical quantum sensors. NV centers in diamond are defects in the diamond lattice that can host quantum states even at room temperature, allowing for highly sensitive magnetic and electric field sensing in biologically relevant environments. This has opened up exciting opportunities in biomedical diagnostics and even in-vivo measurements at the cellular level.

1.1.5 Quantum Sensors in the Future Landscape

The future of quantum sensors is deeply intertwined with the broader field of quantum technologies, which includes quantum computing, quantum communication, and quantum cryptography. As these technologies mature, they are likely to complement and accelerate the development of quantum sensors. For instance, advances in quantum computing may lead to better algorithms for signal processing in quantum sensing, while secure quantum communication networks could benefit from quantum sensors for synchronization and stability. Governments and private enterprises around the world are investing heavily in quantum sensing research, recognizing its transformative potential. National programs, such as the U.S. National Quantum Initiative, the European Quantum Flagship, and similar initiatives in China, Canada, and India, are actively funding the development of next-generation quantum sensors. These efforts aim to bring laboratory-grade sensitivity to real-world environments, paving the way for revolutionary changes in healthcare, environmental monitoring, defense, and space exploration.

1.2 THE PRINCIPLES OF QUANTUM SENSING

1.2.1 Superposition and Entanglement

Quantum sensing is a revolutionary field that utilizes quantum mechanical phenomena to achieve highly sensitive and accurate measurements of physical parameters such as time, magnetic fields, temperature, acceleration, and gravity. Among the most critical principles that underlie quantum sensing are superposition and entanglement. These concepts are cornerstones of quantum mechanics and provide the fundamental advantages that distinguish quantum sensors from their classical counterparts. Superposition and entanglement are not just abstract ideas from quantum theory—they are practical tools that are reshaping the landscape of sensing technologies.

These principles enable quantum sensors to achieve sensitivity and precision far beyond classical limits, opening new frontiers in scientific measurement and real-world applications. While there are still technical challenges to overcome, the rapid progress in quantum engineering and experimental techniques suggests that the widespread use of quantum sensors based on superposition and entanglement is not only possible but imminent. As we move into an era of quantum technologies, these principles will be central to how we observe, understand, and interact with the world around us.

The Principle of Superposition in Quantum Systems

Superposition is one of the defining characteristics of quantum mechanics. It refers to the ability of a quantum system to exist in multiple states simultaneously until it is measured. In classical physics, an object exists in one specific state at a time. For example, a classical particle can be either here or there, but not both. In contrast, a quantum particle such as an electron or an atom can be in a combination of states—both here and there—at the same time. This property allows quantum systems to carry and process more information and react to external stimuli with remarkable precision.

In quantum sensing, superposition is used to enhance the sensitivity of a system to external changes. A common application involves placing atoms in a superposition of energy states using laser pulses. These atoms are then allowed to evolve under the influence of a physical field, such as a magnetic or gravitational field. The superposed states accumulate phase differences based on the field's strength. By measuring these phase shifts, one can infer the magnitude of the physical quantity being probed. This technique is particularly useful in atomic clocks and interferometers, where the difference in phase between two superposed paths reveals extremely accurate measurements.

The ability of superposition to provide enhanced sensitivity is directly related to the coherent nature of quantum systems. The longer a system can maintain coherence—that is, remain in a superposed state without interference from its surroundings—the more precise the measurement can be. This is why quantum sensors often operate in controlled environments like vacuum chambers or at very low temperatures to reduce decoherence. Researchers also work on techniques such as dynamical decoupling and quantum error correction to preserve coherence for longer durations.

Entanglement and Its Role in Enhanced Measurement

Entanglement is another uniquely quantum phenomenon where two or more particles become so deeply connected that the state of one instantaneously influences the state of the other, regardless of the distance between them. This non-local correlation defies classical intuition and has been confirmed by numerous experiments over the decades. In the context of quantum sensing, entanglement is harnessed to improve both sensitivity and precision.

One of the most remarkable aspects of entanglement is that it can be used to create quantum-enhanced sensors that outperform classical limits. For example, in a standard sensor, the precision of a measurement scales with the square root of the number of particles involved (this is known as the shot-noise limit). However, when particles are entangled, this scaling can improve to the Heisenberg limit, where precision increases linearly with the number of particles. This significant improvement allows entangled quantum sensors to detect extremely weak signals that would be undetectable using classical means.

An application of entanglement in sensing can be seen in quantum interferometry, where entangled photons or atoms are used in a Mach-Zehnder or Ramsey interferometer. When entangled particles traverse different paths influenced by a physical quantity such as gravitational waves or electromagnetic fields, the correlations between their states can be used to extract high-resolution information. These devices have applications in gravitational wave detection, dark matter searches, and high-resolution imaging.

Another area where entanglement is particularly useful is in quantum networks of sensors. When multiple sensors are entangled, they can be used collectively to enhance spatial resolution or to cover large geographic areas. For example, entangled atomic clocks located in different places can synchronize more accurately than separate, unentangled clocks. This principle has enormous implications for technologies like GPS, telecommunications, and geophysical exploration.

Combining Superposition and Entanglement for Maximum Sensitivity

While both superposition and entanglement independently offer powerful capabilities, their combined use results in quantum sensors with unmatched sensitivity and precision. A classic example is found in atomic interferometry, where atoms are placed in superposed states and simultaneously entangled with one another. The interferometer measures the interference pattern resulting from phase changes due to external fields, and the entanglement between atoms boosts the resolution of the interference fringes.

The synergy of these principles is not limited to fundamental research; it has practical applications in areas such as magnetometry, gravimetry, and biomedical diagnostics. In quantum magnetometry, for instance, atoms in superposition states are entangled and used to detect minute variations in magnetic fields, such as those produced by neural activity in the brain. This technique offers a non-invasive and ultra-sensitive approach to medical imaging that could revolutionize how neurological disorders are diagnosed and monitored.

Challenges in Implementing Superposition and Entanglement

Despite their theoretical advantages, leveraging superposition and entanglement in practical quantum sensors is technically challenging. One of the major hurdles is decoherence, which is the loss of quantum properties due to interactions with the external environment. Quantum systems are highly sensitive not just to the target signals but also to noise, which can collapse superpositions and destroy entanglement. Researchers must therefore isolate quantum systems and use sophisticated shielding, error correction, and cooling technologies to preserve quantum coherence.

Another challenge is the generation and control of entanglement among a large number of particles. While entangling two particles is relatively straightforward, scaling this up to tens, hundreds, or thousands of particles without introducing noise is a complex task. Advances in quantum optics, ion traps, and solid-state systems are gradually overcoming these barriers, making the production and manipulation of entangled states more practical and reliable.

There are also engineering challenges associated with miniaturization and field deployment. Most high-precision quantum sensors are currently confined to laboratory settings. Making them portable and robust enough for use in harsh environments, such as underground exploration or space missions, requires significant technological innovation in materials science, cryogenics, and vacuum engineering.

Future Prospects of Quantum Sensing with Superposition and Entanglement

The future of quantum sensing is exceedingly promising. As research into quantum systems matures and as our ability to control superposition and entanglement improves, quantum sensors will become more practical, versatile, and powerful. Emerging technologies such as quantum-enhanced lidar, entangled photon spectroscopy, and quantum accelerometers are already in the prototype stage and hint at the vast range of applications waiting to be unlocked.

Governments and industries around the world are heavily investing in quantum technologies, recognizing their transformative potential. Quantum sensing is expected to play a critical role in fields such as environmental monitoring, defense, aerospace, and healthcare. For example, ultra-sensitive quantum gravimeters may help in predicting earthquakes by detecting subtle changes in the Earth's crust, while quantum-enhanced microscopes could offer unprecedented views of molecular structures inside living cells.

In addition, the integration of quantum sensing with other quantum technologies, such as quantum communication and computing, is likely to produce synergistic effects. Quantum networks could synchronize sensors over large distances, and quantum processors could analyze sensor data in ways classical computers cannot match. These interconnected systems could lead to new scientific discoveries and technological breakthroughs that are currently beyond our imagination.

1.2.2 Quantum Interference and Measurement

Quantum sensing is a transformative area of science that leverages the peculiar properties of quantum mechanics to measure physical quantities with exceptional precision. Among the foundational principles of quantum sensing are quantum interference and quantum measurement. These concepts lie at the heart of how quantum systems are manipulated to detect minute changes in the environment, making them indispensable in applications ranging from timekeeping and navigation to gravitational wave detection and biomedical imaging. Quantum interference and measurement are central to the functionality and performance of quantum sensors. By exploiting the wave-like nature of quantum systems and the peculiarities of quantum measurement, researchers can build devices that detect extremely small changes in physical parameters with remarkable precision. Although significant challenges remain in terms of decoherence, measurement fidelity, and practical implementation, the progress in this field is rapid and promising. As our mastery of these quantum principles continues to grow, so too will the range and impact of quantum sensing technologies across science, industry, and daily life.

Understanding Quantum Interference in Sensing

Quantum interference arises from the wave-like nature of particles in quantum mechanics. Unlike classical particles, quantum particles such as photons, electrons, and atoms exhibit wave behavior, meaning that their probability amplitudes can superimpose and interfere with one another. This interference can be constructive or destructive depending on the phase relationship between the waves. When two or more quantum paths combine, their amplitudes interfere, producing patterns that reveal the physical parameters influencing the system.

In quantum sensing, interference is utilized to measure small differences in phase that arise when a quantum system experiences different environmental conditions along different paths. One of the most well-known implementations of quantum interference is the interferometer. Devices such as the Mach-Zehnder, Michelson, and Ramsey interferometers exploit quantum interference by splitting particles into two paths, exposing them to different influences such as magnetic fields, gravity, or acceleration, and then recombining them. The resulting interference pattern encodes the phase differences accumulated along the paths, which can be translated into highly accurate measurements of the external influences.

For example, in atomic interferometry, atoms are cooled to near absolute zero and placed in a superposition of momentum states using laser pulses. These atoms then travel along two separate paths under the influence of gravity or another field. Upon recombination, the phase difference between the two paths causes interference, which is measured to determine the field's strength or direction. This technique is used in applications such as gravimetry, inertial navigation, and even in testing the fundamentals of general relativity.

The key advantage of quantum interference in sensing is its extreme sensitivity to phase changes. Even a tiny perturbation in the environment can alter the interference pattern, making it possible to detect weak signals that are beyond the reach of classical instruments. This sensitivity is directly related to the coherence and purity of the quantum states involved, requiring meticulous control over experimental conditions.

The Role of Quantum Measurement in Sensing

Quantum measurement is another fundamental principle in quantum sensing. Unlike classical measurements, which can be performed without disturbing the system, quantum measurements generally collapse the system's wavefunction, forcing it into one of the eigenstates of the observable being measured. This collapse is an intrinsic part of the quantum formalism and plays a critical role in extracting information from quantum systems.

In the context of quantum sensing, measurement is not merely a passive observation but an active process that determines the outcome of an experiment. The design of the measurement apparatus and the choice of observable significantly affect what information can be extracted and how accurately it can be known. For instance, in quantum magnetometry, the spin state of an atom might be measured to infer the magnetic field's strength. The interaction between the atomic spin and the magnetic field changes the quantum state, and measurement reveals the magnitude of this interaction.

Quantum measurements are often probabilistic, meaning that repeated measurements on identically prepared systems may yield different outcomes. However, by performing many measurements and analyzing the statistical distribution of results, highly precise estimates of physical quantities can be obtained. This probabilistic nature does not hinder quantum sensing but rather complements it, especially when combined with quantum statistical methods and signal processing techniques. Moreover, the process of measurement can itself be engineered to enhance sensitivity. Techniques such as quantum non-demolition measurements allow observation of a quantum system without destroying its coherence, thereby enabling repeated or continuous monitoring. Weak measurements and adaptive measurement protocols are also employed to optimize information extraction without causing significant disruption to the system's state.

Interplay Between Interference and Measurement

The real power of quantum sensing lies in the interplay between interference and measurement. Quantum interference sets the stage by encoding environmental information into the phase and amplitude of quantum states, while quantum measurement extracts that information in a meaningful and interpretable form. Together, these principles allow quantum sensors to reach sensitivity levels that surpass the classical limits defined by the standard quantum limit.

In a typical sensing application, a quantum system is prepared in a coherent superposition, allowed to evolve under the influence of the quantity to be measured, and then subjected to an interference process that reveals phase differences. This is followed by a carefully designed measurement that collapses the system and records the result. The cycle is repeated many times to reduce uncertainty and increase confidence in the measurement outcome.

One illustrative example is the use of nitrogen-vacancy (NV) centers in diamond as quantum sensors. NV centers are defects in the diamond lattice that behave like artificial atoms with quantum states that are sensitive to electric and magnetic fields. By optically initializing and reading out the spin states of NV centers, researchers can use interference between spin states to detect field variations at the nanoscale. The precise control over interference and measurement in these systems enables high-resolution magnetic imaging for materials science and biological research.

Challenges in Quantum Interference and Measurement

Despite their theoretical elegance and experimental utility, implementing interference and measurement in practical quantum sensors poses significant challenges. One major obstacle is decoherence, which refers to the loss of quantum coherence due to interactions with the surrounding environment. Decoherence destroys the phase relationships that are crucial for interference and can render measurements noisy or meaningless. Overcoming decoherence requires isolation, low temperatures, and advanced error correction techniques. Another challenge is the efficiency and fidelity of quantum measurements. Since measurement outcomes are inherently probabilistic and can be influenced by imperfections in the detection system, achieving high precision often necessitates advanced signal processing, calibration, and error mitigation strategies. Additionally, the scalability of quantum interference setups—especially when dealing with multiple particles or complex systems—remains a technical hurdle.

Instrumentation and control systems must also evolve to support quantum sensing technologies. This includes the development of stable lasers, ultra-precise timing systems, and high-performance data acquisition hardware. Integrating quantum sensors into field-deployable or compact devices continues to be an area of intense research and innovation.

Applications and Future Directions

The principles of quantum interference and measurement are already enabling cutting-edge applications across various domains. In navigation, quantum accelerometers and gyroscopes offer drift-free operation, which could replace GPS in environments where satellite signals are unavailable. In timekeeping, optical atomic clocks based on interference principles provide unprecedented precision, facilitating new tests of fundamental physics. In medical imaging, quantum sensors hold the promise of non-invasive, real-time diagnostics with high spatial resolution.

As quantum technologies mature, we can expect even more sophisticated sensors that combine interference and measurement with other quantum resources such as entanglement and squeezing. These hybrid systems could unlock new frontiers in Earth observation, subterranean exploration, and secure communications. Moreover, as quantum sensing becomes integrated with quantum networks and quantum computing, the synergies among these fields may lead to unprecedented scientific discoveries and practical capabilities.

1.3 CLASSICAL VS. QUANTUM SENSORS

In the realm of measurement and detection technologies, sensors play a pivotal role in gathering information about the physical world. Classical sensors, developed and refined over the past century, are widely used in applications ranging from environmental monitoring and navigation to healthcare and industrial automation. However, the advent of quantum technologies has introduced a new class of sensors—quantum sensors—that promise unprecedented precision and sensitivity. The comparison between classical and quantum sensors is not merely one of technical specifications but of underlying physical principles. Classical and quantum sensors represent two different paradigms in the measurement of physical quantities. Classical sensors rely on established physical principles and are highly practical, cost-effective, and widely used. Quantum sensors, while more complex and currently limited in availability, offer extraordinary sensitivity and open the door to measurements that were previously thought impossible. The continued development of quantum technologies, combined with the proven reliability of classical systems, suggests a future in which both types of sensors coexist, each playing a crucial role in advancing science, industry, and everyday life. As the boundaries of what can be measured continue to expand, the interplay between classical and quantum sensors will be at the forefront of technological progress.

1.3.1 Foundations of Classical Sensing

Classical sensors operate based on the principles of classical physics. They rely on the predictable and deterministic behavior of macroscopic objects and electromagnetic fields. For instance, a traditional thermometer measures temperature by observing the expansion of mercury or a change in electrical resistance. Similarly, accelerometers measure changes in motion using mechanical or piezoelectric components. Classical sensors function well under most conditions and have been engineered for robustness, repeatability, and integration into electronic systems. The precision of classical sensors is ultimately limited by factors such as thermal noise, mechanical stability, and electronic interference. These limitations define what is known as the "standard quantum limit" in precision measurements. Though they are not based on quantum mechanics, classical sensors can still achieve remarkable accuracy within this boundary, especially when enhanced through engineering techniques such as feedback control and signal amplification.

1.3.2 Principles of Quantum Sensing

Quantum sensors, on the other hand, derive their capabilities from uniquely quantum phenomena such as superposition, entanglement, and quantum interference. Instead of measuring physical quantities through macroscopic changes, quantum sensors probe microscopic quantum states of atoms, ions, or photons. These states are exquisitely sensitive to external influences, allowing quantum sensors to detect minute changes in parameters like magnetic fields, gravitational forces, temperature, and time. For example, an atomic clock is a type of quantum sensor that uses the precise frequency of microwave or optical transitions in atoms to keep time with extraordinary accuracy. Similarly, nitrogen-vacancy centers in diamonds can detect magnetic fields at the nanoscale by monitoring shifts in spin states. These sensors take advantage of the fact that quantum systems accumulate phase information in a way that is directly influenced by their environment, and this information can be extracted through quantum measurements.

1.3.3 Sensitivity and Precision

One of the most significant distinctions between classical and quantum sensors lies in their sensitivity. Quantum sensors can surpass the standard quantum limit that constrains classical devices. This is made possible by exploiting quantum entanglement and squeezing, which reduce the uncertainty in one measurement variable at the expense of another, thereby increasing measurement precision beyond classical limits. For instance, interferometric gravitational wave detectors like LIGO use squeezed light—a quantum-enhanced state of light—to reduce noise and improve sensitivity. Classical sensors, limited by shot noise and thermal fluctuations, cannot achieve such levels of detection for extremely weak signals. Similarly, quantum magnetometers can detect changes in magnetic fields down to femtotesla levels, a sensitivity unachievable by traditional Hall-effect sensors or fluxgate magnetometers.

1.3.4 Decoherence and Stability

Despite their superior sensitivity, quantum sensors face challenges related to decoherence and environmental noise. Quantum systems are inherently fragile; they can easily lose their coherence through interactions with their surroundings. This makes maintaining stable quantum states over time and space a major hurdle in practical quantum sensing. Ensuring low temperatures, isolation from electromagnetic disturbances, and advanced error correction methods are essential to preserve quantum advantages. Classical sensors, by contrast, are generally more robust and reliable in harsh or uncontrolled environments. They do not require the extreme conditions often needed for quantum systems to function, such as cryogenic cooling or ultra-high vacuum. This gives classical sensors a practical edge in many real-world applications where ease of use and durability are paramount.

1.3.5 Size and Scalability

Another area where classical and quantum sensors differ is in their scalability and size. Classical sensors are often compact, energy-efficient, and easily mass-produced. They can be integrated into mobile devices, vehicles, and wearables, making them ideal for widespread deployment. Quantum sensors, especially those in early developmental stages, tend to be bulkier and more complex due to the equipment needed to manipulate and read quantum states. However, recent advances in quantum technologies have led to miniaturized quantum sensors, such as chip-scale atomic clocks and portable quantum gravimeters. The field is rapidly progressing toward more scalable and user-friendly designs, potentially leading to a future where quantum sensors are as ubiquitous as their classical counterparts.

1.3.6 Application Domains

Classical sensors are deeply embedded in everyday technologies. They are used in smartphones to detect orientation and movement, in cars to monitor tire pressure and braking, in weather stations to measure temperature and humidity, and in medical devices to track vital signs. Their versatility, low cost, and mature manufacturing infrastructure make them suitable for a wide array of applications. Quantum sensors, although more specialized, open up possibilities in domains that demand ultra-high precision. In navigation, quantum accelerometers and gyroscopes can operate without reliance on GPS signals, providing accurate positioning in underground or space environments. In medicine, quantum sensors promise non-invasive brain imaging and early disease detection through sensitive biomagnetic measurements. In scientific research, they enable tests of fundamental physics, including the search for dark matter and the detection of gravitational waves.

1.3.7 Cost and Accessibility

Currently, classical sensors hold a clear advantage in terms of cost and accessibility. They benefit from decades of industrial refinement, mass production, and standardization. Quantum sensors, on the other hand, involve cutting-edge technology that is often expensive and limited to research laboratories or specialized industries. However, as quantum technologies mature and production techniques improve, the cost of quantum sensors is expected to decrease. Governments and private companies around the world are investing heavily in quantum research, anticipating a future where quantum sensors become more economically viable and widely available.

1.3.8 Future Outlook

The comparison between classical and quantum sensors is not about replacing one with the other but about understanding their complementary strengths. Classical sensors will continue to dominate in applications requiring reliability, affordability, and ease of integration. Quantum sensors will thrive in scenarios where ultimate sensitivity, accuracy, and novel capabilities are required. Hybrid systems that combine classical and quantum elements are also on the horizon. For example, classical electronics can be used to control and stabilize quantum sensors, while quantum-enhanced data can improve the performance of classical systems. This convergence of classical and quantum technologies is likely to shape the next wave of innovations in sensing and measurement.

1.4 KEY ADVANTAGES OF QUANTUM SENSING

Quantum sensing represents one of the most promising frontiers in modern technology, offering measurement capabilities that far exceed those of traditional classical sensors. As quantum physics continues to move from theoretical research into practical application, quantum sensing stands out as a transformative innovation. By leveraging fundamental quantum phenomena such as superposition, entanglement, and quantum tunneling, these sensors provide unmatched precision, sensitivity, and resolution. The implications are vast, touching fields ranging from healthcare and navigation to environmental monitoring and fundamental physics research. Quantum sensing stands at the intersection of scientific innovation and practical utility. With advantages that include unparalleled sensitivity, subatomic-scale resolution, enhanced signal-to-noise ratios, and resilience in extreme environments, quantum sensors are set to redefine how we observe and interact with the physical world. Their applications span from precision timekeeping and medical diagnostics to navigation, space exploration, and fundamental physics research. As these sensors continue to evolve and transition from laboratories to commercial markets, their impact will be felt across diverse sectors. The future of sensing is inherently quantum, promising a new era of measurement and discovery that was once thought impossible.

1.4.1 Unparalleled Sensitivity

One of the primary advantages of quantum sensors is their extraordinary sensitivity to external changes. Quantum systems are inherently more responsive to their environment due to the delicate nature of quantum states. For instance, a quantum sensor based on atomic spin states can detect minute fluctuations in magnetic or electric fields that are orders of magnitude weaker than those detectable by classical sensors. This heightened sensitivity is made possible by the quantum coherence of the system, allowing the sensor to accumulate phase shifts or frequency shifts with extreme accuracy. In practical terms, this translates to the ability to measure very small changes in position, force, field strength, or time with unprecedented precision.

1.4.2 Subatomic-Scale Resolution

Quantum sensors also offer spatial resolution capabilities at the subatomic level. In contrast to classical sensors, which are constrained by noise and material limitations, quantum sensors can detect physical properties at scales previously inaccessible. For example, nitrogen-vacancy (NV) centers in diamond can probe magnetic fields at the level of individual molecules or even atoms. This ultra-high spatial resolution is invaluable in areas such as materials science, nanotechnology, and biological imaging. Scientists can use quantum sensors to observe interactions at the cellular or molecular level without disturbing the system, enabling a new class of non-invasive, high-resolution diagnostics and experiments.

1.4.3 Improved Timekeeping and Frequency Standards

Atomic clocks, a type of quantum sensor, illustrate one of the clearest applications of quantum precision. By exploiting the natural oscillation frequencies of atoms such as cesium or ytterbium, atomic clocks provide the most accurate timekeeping devices available. The consistency of atomic transitions ensures that quantum clocks lose only a fraction of a second over millions of years. This advantage has profound implications for technologies that rely on precise timing, including GPS navigation, telecommunications, and financial systems. Quantum-based time standards also enhance synchronization across global infrastructures, enabling more efficient and coordinated operations in a variety of sectors.

1.4.4 Reduced Systematic Errors through Quantum Interference

Another key advantage of quantum sensing is the reduction of systematic errors through the use of quantum interference. Interference arises when quantum particles such as atoms or photons exist in a superposition of states and their probability amplitudes interfere constructively or destructively.

In a quantum sensor, this effect is harnessed to compare paths or states with incredible fidelity. Atom interferometers, for example, use laser pulses to split and recombine atomic wavefunctions, allowing them to detect gravitational gradients, accelerations, and rotations with remarkable clarity. The interference pattern that results is highly sensitive to external perturbations, leading to measurements with low noise and high repeatability.

1.4.5 Enhanced Signal-to-Noise Ratio

Quantum sensors can achieve a significantly enhanced signal-to-noise ratio (SNR) compared to classical alternatives. This improvement is often achieved by utilizing quantum entanglement or squeezing techniques that reduce uncertainty in one variable while allowing greater uncertainty in its conjugate variable, in accordance with Heisenberg's uncertainty principle. For instance, in optical sensing, squeezed light can be used to reduce quantum noise and improve detection accuracy in low-light conditions. This capability is especially valuable in applications where signals are weak or buried under background noise, such as astronomical observations, detection of gravitational waves, or scanning tunneling microscopy.

1.4.6 Capability in Extreme Environments

Quantum sensors have demonstrated resilience and performance in extreme or challenging environments where classical sensors may fail. This includes deep-sea conditions, outer space, underground locations, and high-radiation zones. Since some quantum sensors operate at the atomic or subatomic level, they can function in isolated quantum states that are not easily disrupted by environmental conditions. For example, cold atom gravimeters can operate effectively in space missions to map Earth's gravity or monitor planetary surfaces. Similarly, quantum magnetometers can detect mineral deposits or voids in geological structures without needing invasive drilling, making them useful in archaeology, mining, and civil engineering.

1.4.7 GPS-Independent Navigation Systems

One of the emerging applications of quantum sensing is in navigation systems that do not rely on external signals like GPS. Quantum accelerometers and gyroscopes can measure inertial changes with extreme precision, allowing for accurate dead-reckoning navigation over long distances. This advantage is crucial in environments where GPS signals are unavailable or unreliable, such as underwater, underground, or in military operations that require stealth. These sensors can maintain navigational accuracy without recalibration for extended periods, providing a reliable alternative to satellite-based positioning systems.

1.4.8 Advancing Fundamental Physics

Quantum sensors are also powerful tools for probing the fundamental laws of physics. They are being used in experiments to detect gravitational waves, test the equivalence principle, measure the fine-structure constant, and search for dark matter. The extreme sensitivity of quantum sensors enables researchers to detect tiny anomalies that could indicate new physics beyond the Standard Model. For instance, atomic interferometers can sense extremely weak forces or measure minute variations in space-time curvature. In this sense, quantum sensing is not only a technological advancement but also a vital instrument in advancing our understanding of the universe.

1.4.9 Non-Invasive Medical Diagnostics

In the field of medicine, quantum sensors are paving the way for new forms of non-invasive diagnostics. Quantum magnetometers, such as those based on NV centers in diamond or superconducting quantum interference devices (SQUIDs), can detect magnetic fields produced by neuronal activity in the brain or by the beating of the heart. These techniques allow for functional imaging of organs without the need for radiation or invasive procedures. This could lead to safer, more accurate diagnostics for conditions such as epilepsy, cardiac arrhythmia, or neurodegenerative disorders, greatly enhancing patient care and early disease detection.

1.4.10 Long-Term Economic and Strategic Value

From a broader perspective, the strategic value of quantum sensing extends to economic and national security domains. Countries that invest in quantum technologies are likely to gain competitive advantages in defense, telecommunications, energy, and transportation. Quantum sensors can be used in submarine detection, secure navigation, infrastructure monitoring, and even espionage. Their potential to disrupt current technologies means they are rapidly becoming a focus of national quantum initiatives around the world. As the cost of quantum components continues to fall and integration improves, the economic impact of quantum sensing is expected to be profound and widespread.

1.5 HISTORY AND EVOLUTION OF QUANTUM SENSING TECHNOLOGIES

The journey of quantum sensing technologies traces its roots to the early 20th century when the foundations of quantum mechanics were first laid. Initially, the domain of quantum physics remained largely theoretical, explored by pioneers such as Max Planck, Niels Bohr, Albert Einstein, and Werner Heisenberg. Their contributions to the understanding of quantum phenomena like quantization of energy, the uncertainty principle, and wave-particle duality set the stage for future technological applications. While the primary interest at the time was in explaining atomic and subatomic behavior, these theories eventually laid the groundwork for practical innovations. Among these, quantum sensing emerged as a distinct field capable of utilizing the subtle and powerful behaviors of quantum systems for precise measurement purposes. The history and evolution of quantum sensing technologies reflect a remarkable journey from fundamental physics to groundbreaking applications. What began as a theoretical exploration of atomic and subatomic phenomena has now matured into a field that offers revolutionary capabilities for measuring the physical world. The path from atomic clocks and SQUIDs to NV centers and cold atom interferometers showcases how technological progress, cross-disciplinary collaboration, and strategic investment have driven innovation. As we stand on the cusp of a quantum era, quantum sensors are poised to play a central role in shaping scientific discovery, industrial transformation, and societal advancement.

During the mid-20th century, as experimental physics advanced, scientists began to manipulate quantum systems with increasing control. The invention of the laser in 1960 was a pivotal moment. Lasers allowed for highly controlled interaction with atoms and photons, a necessity for any quantum measurement. This control made it possible to explore phenomena such as quantum interference and superposition in laboratory conditions. At the same time, technologies such as atomic clocks were beginning to showcase the potential of quantum systems in real-world applications. The first caesium atomic clock was built in 1955, introducing an entirely new standard of timekeeping based on the frequency of atomic transitions. This was one of the earliest examples of a practical quantum sensor.

In the 1970s and 1980s, the development of superconducting quantum interference devices (SQUIDs) marked a significant milestone. SQUIDs, which detect minute changes in magnetic fields by exploiting quantum interference in superconductors, became crucial tools in both scientific research and applied technologies. These devices represented one of the first widely recognized and used quantum sensors. They were used in fields such as geophysics, medicine (e.g., magnetoencephalography for brain activity mapping), and material science. The precision offered by SQUIDs illustrated the transformative potential of quantum sensing in disciplines where classical sensors had reached their limitations.

The field gained momentum in the 1990s with breakthroughs in laser cooling and trapping of atoms. Physicists learned how to cool atoms to near absolute zero and trap them using magnetic and optical fields. These techniques allowed for the creation of Bose-Einstein condensates and opened up unprecedented possibilities for quantum manipulation. With ultracold atoms, scientists could build atom interferometers, devices that use wave-like properties of atoms to make ultra-precise measurements. These interferometers became essential for measuring gravitational fields, inertial forces, and time, making them valuable tools in both fundamental research and applied engineering. The use of cold atoms was a major turning point in the evolution of quantum sensing.

Entering the 21st century, quantum sensing began to emerge as a distinct research and application area, benefitting from the broader revolution in quantum technology. Advancements in quantum optics, quantum information theory, and nanotechnology provided a rich landscape for the development of novel sensors. In particular, nitrogen-vacancy (NV) centers in diamond became a focus of interest. These are defects in the diamond crystal lattice where a nitrogen atom replaces a carbon atom next to a vacancy. NV centers exhibit quantum properties that can be manipulated at room temperature, allowing for sensitive magnetic field detection with nanometer-scale resolution. Their compatibility with biological environments made them especially attractive for medical diagnostics and neuroscience research.

The rise of quantum technologies in the 2010s, driven by government initiatives and increasing private investment, accelerated the evolution of quantum sensing. Programs such as the U.S. National Quantum Initiative, the EU Quantum Flagship, and China's National Key R&D Program placed quantum sensing alongside computing and communication as strategic priorities. As a result, numerous academic and commercial efforts emerged to transition quantum sensors from laboratory experiments to deployable technologies. Innovations in solid-state systems, photonics, and chip-scale integration have allowed for miniaturization and robustness, crucial for deploying quantum sensors in real-world settings such as satellites, submarines, or hospitals.

Recent years have seen the application of quantum sensors in a variety of fields with increasing sophistication. In geology and civil engineering, gravimeters based on quantum principles are being used to map underground structures or monitor volcanoes. In space exploration, quantum accelerometers and gyroscopes offer GPS-independent navigation systems that are being tested on satellites. In healthcare, magnetometers using quantum effects are enabling high-resolution brain imaging without the need for cryogenic cooling. Furthermore, quantum sensors are playing a growing role in defense and security, providing stealthy detection capabilities, precise timing for communication networks, and tools for submarine tracking.

Quantum sensors are also helping to answer profound scientific questions. Atom interferometers are being used to detect gravitational waves and to probe the nature of dark energy. Quantum-enhanced telescopes may soon be able to detect exoplanets with greater accuracy or observe cosmic phenomena at scales previously unattainable. Moreover, these sensors are being explored for their role in testing the fundamental principles of physics, such as the equivalence principle in general relativity or searching for time variations in fundamental constants.

One of the most important aspects of the recent evolution of quantum sensing is the move toward practical implementation and commercial deployment. Startups and established tech companies alike are investing in quantum sensing, developing ruggedized and cost-effective devices for industry use. Examples include portable atomic clocks, handheld magnetic sensors, and gravity sensors that can be mounted on drones. These innovations signify a shift from quantum sensing as a niche scientific tool to a mainstream technological solution capable of transforming industries. Fields such as autonomous vehicles, environmental monitoring, precision agriculture, and even financial systems are beginning to explore how quantum sensing might provide competitive advantages.

In the future, the trajectory of quantum sensing technologies is likely to continue accelerating as quantum hardware becomes more reliable, scalable, and affordable. Integration with artificial intelligence and machine learning is also on the horizon, enabling smarter interpretation of quantum sensor data in real time. Efforts are underway to build hybrid systems that combine classical and quantum sensors for enhanced performance and redundancy. Additionally, as quantum communication networks mature, there may be opportunities for distributed quantum sensors, enabling collaborative sensing across vast distances with entangled systems.

CHAPTER 2

Fundamental Concepts in Quantum Sensing

2.1 QUBITS AND QUANTUM STATES IN SENSING

2.1.1 Qubit

A qubit or quantum bit is a basic unit of quantum information—the quantum version of the classic binary bit physically realized with a two-state device. A qubit is a two-state (or two-level) quantum-mechanical system, one of the simplest quantum systems displaying the peculiarity of quantum mechanics. Examples include the spin of the electron in which the two levels can be taken as spin up and spin down; or the polarization of a single photon in which the two spin states (left-handed and the right-handed circular polarization) can also be measured as horizontal and vertical linear polarization. In a classical system, a bit would have to be in one state or the other. However, quantum mechanics allows the qubit to be in a coherent superposition of multiple states simultaneously, a property that is fundamental to quantum mechanics and quantum computing.

Bit versus Qubit

A binary digit, characterized as 0 or 1, is used to represent information in classical computers. When averaged over both of its states (0,1), a binary digit can represent up to one bit of information content, where a bit is the basic unit of information.

In classical computer technologies, a *processed* bit is implemented by one of two levels of low direct current voltage, and whilst switching from one of these two levels to the other, a so-called "forbidden zone" between two logic levels must be passed as fast as possible, as electrical voltage cannot change from one level to another instantly.

There are two possible outcomes for the measurement of a qubit—usually taken to have the value "0" and "1", like a bit. However, whereas the state of a bit can only be binary (either 0 or 1), the general state of a qubit according to quantum mechanics can arbitrarily be a coherent superposition of *all* computable states simultaneously. Moreover, whereas a measurement of a classical bit would not disturb its state, a measurement of a qubit would destroy its coherence and irrevocably disturb the superposition state. It is possible to fully encode one bit in one qubit. However, a qubit can hold more information, e.g., up to two bits using superdense coding. A bit is always completely in either one of its two states, and a set of n bits (e.g. a processor register or some bit array) can only hold a single of its 2^n possible states at any time. A quantum state can be in a superposition state, which means that the qubit can have non-zero probability amplitude in both its states simultaneously (popularly expressed as "it can be in both states simultaneously"). A qubit requires two complex numbers to describe its two probability amplitudes, and these two complex numbers can together be viewed as a 2-dimensional complex vector, which is called a *quantum state vector*, or *superposition state vector.* Alternatively and equivalently, the value stored in a qubit can be described as a single point in a 2-dimensional complex coordinate space.

Furthermore, a set of n bits can be represented by n binary digits, simply by concatenating the representations of each of the bits, whereas a set of n qubits, which is also called a register, requires 2^n complex numbers to describe its superposition state vector. This is because the n qubits are not independent from one another and therefore the register cannot be described by breaking it down and describing the individual qubits.

Standard Representation

In quantum mechanics, the general quantum state of a qubit can be represented by a linear superposition of its two orthonormal basis states (or basis vectors). These vectors are usually denoted as $|0\rangle = \begin{bmatrix} 1 \\ 0 \end{bmatrix} and\, |1\rangle = \begin{bmatrix} 0 \\ 1 \end{bmatrix}$. They are written in the conventional Dirac—or "bra-ket"—notation; the $|0\rangle\, and\, |1\rangle$ are pronounced "ket 0" and "ket 1", respectively. These two orthonormal basis states, $\{|0\rangle, |1\rangle\}$, together called the computational basis, are said to span the two-dimensional linear vector (Hilbert) space of the qubit.

Qubit basis states can also be combined to form product basis states. A set of qubits taken together is called a quantum register. For example, two qubits could be represented in a four-dimensional linear vector space spanned by the following product basis states:

$$|00\rangle = \begin{bmatrix} 1 \\ 0 \\ 0 \\ 0 \end{bmatrix}, |10\rangle = \begin{bmatrix} 0 \\ 0 \\ 1 \\ 0 \end{bmatrix}, |10\rangle = \begin{bmatrix} 0 \\ 0 \\ 1 \\ 0 \end{bmatrix}, and\, |11\rangle = \begin{bmatrix} 0 \\ 0 \\ 0 \\ 1 \end{bmatrix}$$

Qubit States

With a typical mapping $|H\rangle = |0\rangle\, and\, |V\rangle = |1\rangle$, quantum states $(|0\rangle \pm |1\rangle)/\sqrt{2}$ have a direct physical representation, both easily demonstrable experimentally in a class with linear polarizers and, for real $\alpha\, and\, \beta$, matching the high-school definition of orthogonality. A pure qubit state is a coherent superposition of the basis states. This means that a single qubit (ψ) can be described by a linear combination of $|0\rangle\, and\, |1\rangle$:

$$|\psi\rangle = \alpha|0\rangle + \beta|1\rangle$$

where α and β are the probability amplitudes, and are both complex numbers. When we measure this qubit in the standard basis, according to the Born rule, the probability of outcome $|0\rangle$ with value "0" is $|\alpha|^2$ and the probability of outcome $|1\rangle$ with value "1" is $|\beta|^2$. Because the absolute squares of the amplitudes equate to probabilities, it follows that $\alpha\ and\ \beta$ must be constrained according to the second axiom of probability theory by the equation

$$|\alpha|^2 + |\beta|^2 = 1.$$

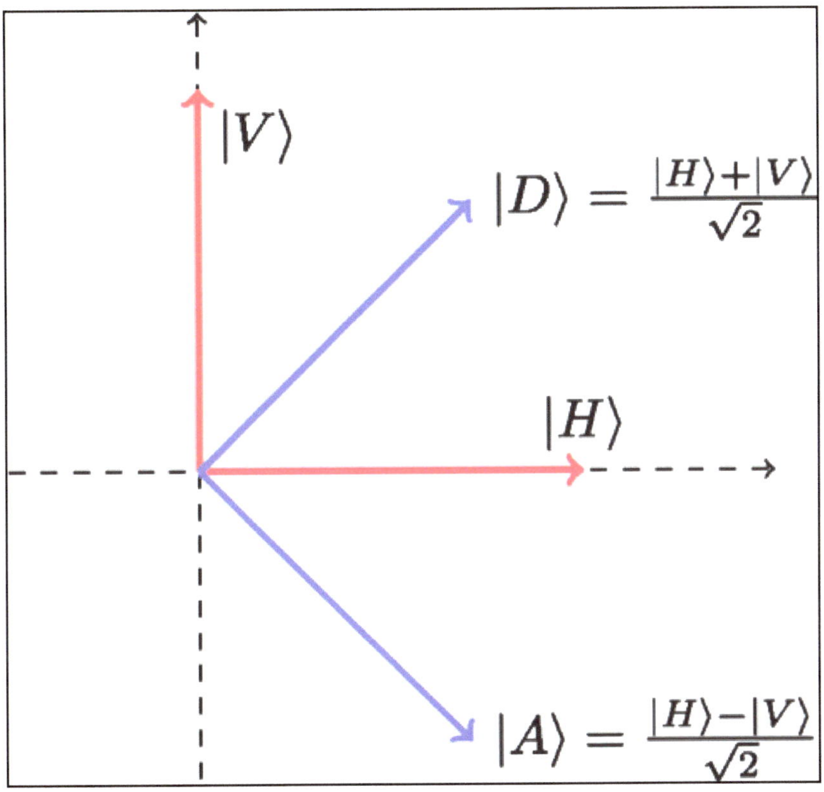

Fig: Polarization of Light Offers a Straightforward Way to Present Orthogonal States

The probability amplitudes, $\alpha\ and\ \beta$, encode more than just the probabilities of the outcomes of a measurement; the *relative phase* between $\alpha\ and\ \beta$ is for example responsible for quantum interference.

Bloch Sphere Representation

The probability amplitudes for the superposition state, $|\psi\rangle = \alpha|0\rangle + \beta|1\rangle$, are given by,

$$\alpha = \cos\left(\frac{\theta}{2}\right) and\ \beta = e^{i\varphi}\sin\left(\frac{\theta}{2}\right)$$

It might, at first sight, seem that there should be four degrees of freedom in $|\psi\rangle = \alpha|0\rangle + \beta|1\rangle$, as α and β are complex numbers with two degrees of freedom each. However, one degree of freedom is removed by the normalization constraint $|\alpha|^2 + |\beta|^2 = 1$. This means, with a suitable change of coordinates,

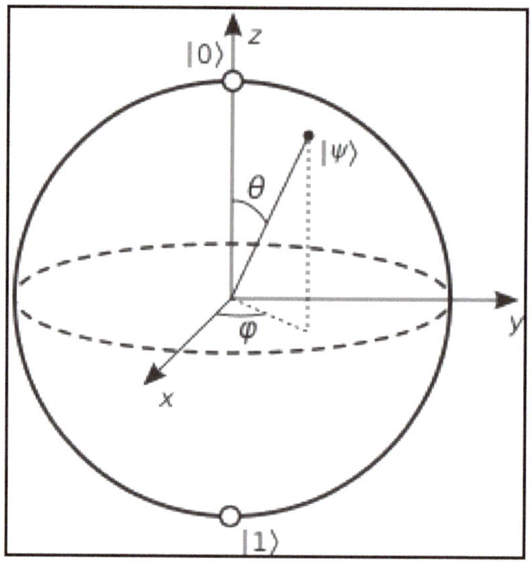

Fig: Bloch Sphere Representation of a Qubit.

one can eliminate one of the degrees of freedom. One possible choice is that of Hopf coordinates:

$$a = e^{i\delta}\cos\frac{\theta}{2},$$

$$\beta = e^{i(\delta+\varphi)}\sin\frac{\theta}{2}.$$

Additionally, for a single qubit the *global phase* of the state e i $e^{i\delta}$ has no physically observable consequences, so we can arbitrarily choose α to be real (or β in the case that α is zero), leaving just two degrees of freedom:

$$\alpha = \cos\frac{\theta}{2},$$

$$\beta = e^{i\varphi}\sin\frac{\theta}{2},$$

The possible quantum states for a single qubit can be visualised using a Bloch sphere. Represented on such a 2-sphere, a classical bit could only be at the "North Pole" or the "South Pole", in the locations where $|0\rangle\, and |1\rangle$ are respectively. This particular choice of the polar axis is arbitrary, however. The rest of the surface of the Bloch sphere is inaccessible to a classical bit, but a pure qubit state can be represented by any point on the surface. For example, the pure qubit state $(|0\rangle + |1\rangle)/\sqrt{2}$ would lie on the equator of the sphere at the positive X-axis. In the classical limit, a qubit, which can have quantum states anywhere on the Bloch sphere, reduces to the classical bit, which can be found only at either poles. The surface of the Bloch sphere is a two-dimensional space, which represents the observable state space of the pure qubit states. This state space has two local degrees of freedom, which can be represented by the two angles $\varphi \, and \, \theta.$

Mixed State

Coherence is essential for a qubit to be in a superposition state. With interactions, quantum noise and decoherence, it is possible to put the qubit in a mixed state, a statistical combination or "incoherent mixture" of different pure states. Mixed states can be represented by points *inside* the Bloch sphere (or in the Bloch ball). A mixed qubit state has three degrees of freedom: the angles $\varphi \, and \, \theta.$, as well as the length r r of the vector that represents the mixed state. Quantum error correction can be used to maintain the purity of qubits.

Operations on Qubits

There are various kinds of physical operations that can be performed on qubits.

- Quantum logic gates, building blocks for a quantum circuit in a quantum computer, operate on a set of qubits (a register); mathematically, the qubits undergo a (reversible) unitary transformation described by multiplying the quantum gates unitary matrix with the quantum state vector. The result from this multiplication is a new quantum state vector.
- Quantum measurement is an irreversible operation in which information is gained about the state of a single qubit, and coherence is lost. The result of the measurement of a single qubit with the state $|\psi\rangle = \alpha|0\rangle + \beta|1\rangle$ will be either $|0\rangle$ with probability $|\alpha|^2 or |1\rangle$ with probability | β | 2 . Measurement of the state of the qubit alters the magnitudes of α and β. For instance, if the result of the measurement is $|\beta|^2$, α is changed to 0 and β is changed to 1, while the phase factor e i $|1\rangle$ is no longer experimentally accessible. If measurement is performed on a qubit that is entangled, the measurement may collapse the state of the other entangled qubits.
- Initialization or re-initialization to a known value, often $e^{i\phi}$. This operation collapses the quantum state (exactly like with measurement). Initialization to $|0\rangle$ may be implemented logically or physically: Logically as a measurement, followed by the application of the Pauli-X gate if the result from the measurement was $|1\rangle$. Physically, for example if it is a superconducting phase qubit, by lowering the energy of the quantum system to its ground state.

- Sending the qubit through a quantum channel to a remote system or machine (an I/O operation), potentially as part of a quantum network.

Quantum Entanglement

An important distinguishing feature between qubits and classical bits is that multiple qubits can exhibit quantum entanglement; the qubit itself is an exhibition of quantum entanglement. In this case, quantum entanglement is a local or nonlocal property of two or more qubits that allows a set of qubits to express higher correlation than is possible in classical systems.

The simplest system to display quantum entanglement is the system of two qubits. Consider, for example, two entangled qubits in the $|\Phi^+\rangle$ Bell state:

$$\frac{1}{\sqrt{2}}(|00\rangle + |11\rangle).$$

Quantum Register

A number of qubits taken together is a qubit register. Quantum computers perform calculations by manipulating qubits within a register.

Qudits and Qutrits

The term qudit denotes the unit of quantum information that can be realized in suitable *d*-level quantum systems. A qubit register that can be measured to *N* states is identical to an *N*-level qudit. A rarely used synonym for qudit is quNit, since both *d* and *N* are frequently used to denote the dimension of a quantum system.

Qudits are similar to the integer types in classical computing, and may be mapped to (or realized by) arrays of qubits. Qudits where the *d*-level system is not an exponent of 2 cannot be mapped to arrays of qubits. It is for example possible to have 5-level qudits.

In 2017, scientists at the National Institute of Scientific Research constructed a pair of qudits with 10 different states each, giving more computational power than 6 qubits.

In 2022, researchers at the University of Innsbruck succeeded in developing a universal qudit quantum processor with trapped ions. In the same year, researchers at Tsinghua University's Center for Quantum Information implemented the dual-type qubit scheme in trapped ion quantum computers using the same ion species. In 2025, the Innsbruck team managed to simulate two-dimensional lattice gauge theories on their qudit quantum computer.

Also in 2022, researchers at the University of California, Berkeley developed a technique to dynamically control the cross-Kerr interactions between fixed-frequency qutrits, achieving high two-qutrit gate fidelities. This was followed by a demonstration of extensible control of superconducting qudits up to $d = 4$ in 2024 based on programmable two-photon interactions.

Similar to the qubit, the qutrit is the unit of quantum information that can be realized in suitable 3-level quantum systems. This is analogous to the unit of classical information trit of ternary computers. Besides the advantage associated with the enlarged computational space, the third qutrit level can be exploited to implement efficient compilation of multi-qubit gates.

Qubit Storage

In 2008 a team of scientists from the U.K. and U.S. reported the first relatively long (1.75 seconds) and coherent transfer of a superposition state in an electron spin "processing" qubit to a nuclear spin "memory" qubit. This event can be considered the first relatively consistent quantum data storage, a vital step towards the development of quantum computing. In 2013, a modification of similar systems (using charged rather than neutral donors) has dramatically extended this time, to 3 hours at very low temperatures and 39 minutes at room temperature. Room temperature preparation of a qubit based on electron spins instead of nuclear spin was also demonstrated by a team of scientists from Switzerland and Australia. An increased coherence of qubits is being explored by researchers who are testing the limitations of a Ge hole spin-orbit qubit structure.

Physical Implementations

Any two-level quantum-mechanical system can be used as a qubit. Multilevel systems can be used as well, if they possess two states that can be effectively decoupled from the rest (e.g., the ground state and first excited state of a nonlinear oscillator). There are various proposals. Several physical implementations that approximate two-level systems to various degrees have been successfully realized. Similarly to a classical bit, where the state of a transistor in a processor, the magnetization of a surface in a hard disk, and the presence of current in a cable can all be used to represent bits in the same computer, an eventual quantum computer is likely to use various combinations of qubits in its design.

All physical implementations are affected by noise. The so-called T_1 lifetime and T_2 dephasing time are a time to characterize the physical implementation and represent their sensitivity to noise. A higher time does not necessarily mean that one or the other qubit is better suited for quantum computing because gate times and fidelities need to be considered, too. Different applications like quantum sensing, quantum computing and quantum communication use different implementations of qubits to suit their application.

The following table is an incomplete list of physical implementations of qubits, and the choices of basis are by convention only.

Physical support	Name	Information support	$\|0\rangle$	$\|1\rangle$
photon	polarization encoding	polarization of light	horizontal	vertical
	number of photons	Fock state	vacuum	single-photon state
	time-bin encoding	time of arrival	early	late
coherent state of light	squeezed light	quadrature	amplitude-squeezed state	phase-squeezed state
electrons	electronic spin	spin	up	down
	electron number	charge	no electron	two electron
nucleus	nuclear spin addressed through NMR	spin	up	down
neutral atom	atomic energy level	spin	up	down
trapped ion	atomic energy level	spin	up	down
Josephson junction	superconducting charge qubit	charge	uncharged superconducting island (Q = 0)	charged superconducting island (Q = 2e, one extra Cooper pair)
	superconducting flux qubit	current	clockwise current	counterclockwise current
	superconducting phase qubit	energy	ground state	first excited state
singly charged quantum dot pair	electron localization	charge	electron on left dot	electron on right dot
quantum dot	dot spin	spin	down	up
gapped topological system	non-abelian anyons	braiding of excitations	depends on specific topological system	depends on specific topological system
vibrational qubit	vibrational states	phonon/vibron	superposition	superposition
van der Waals heterostructure	electron localization	charge	electron on bottom sheet	electron on top sheet

2.1.2 Quantum State

Quantum sensing is an emerging field that leverages the peculiar principles of quantum mechanics to measure physical quantities with unparalleled sensitivity. At the heart of this powerful approach are qubits—quantum bits—which form the foundational units of quantum information. Unlike classical bits that exist in a binary state of either 0 or 1, qubits can exist in a superposition of both states simultaneously. This property, along with entanglement and coherence, makes qubits uniquely suited for applications in quantum sensing. By harnessing quantum properties, sensors using qubits can detect minute variations in magnetic fields, temperature, pressure, and other environmental parameters with much greater precision than classical devices.

Understanding Quantum States

A quantum state is the complete description of a quantum system, encompassing all the measurable information about a particle or group of particles. In the context of sensing, quantum states represent the dynamic condition of qubits as they interact with external stimuli. These states evolve over time and can be precisely manipulated to respond to environmental changes. The ability to prepare, control, and read out specific quantum states is essential to achieving high-precision measurements. For instance, a quantum sensor can be initialized in a specific state and then allowed to evolve under the influence of a magnetic field, with the final state providing insight into the field's strength and direction.

The Power of Superposition in Measurement

Superposition is one of the key quantum phenomena enabling enhanced sensing. When a qubit is in a superposition, it exists in a blend of both its base states simultaneously. This means it can process and react to information in a more nuanced way than a classical bit. In practical terms, when a superposed qubit interacts with its environment, its wavefunction acquires a phase that reflects the influence of external physical parameters. The resulting phase shift can be measured with extremely high precision, allowing researchers to detect subtle changes in forces, fields, or energy levels that would be imperceptible to traditional sensors.

Quantum Coherence and Its Role in Sensing

Another essential feature of qubits is coherence, which refers to the duration a qubit can maintain its quantum state without interference from its environment. Coherence time determines how long a qubit can accumulate information from external influences before losing its quantum properties. Longer coherence times enable more precise measurements, making coherence one of the most important factors in designing effective quantum sensors. Advanced isolation techniques and improved material engineering have significantly enhanced coherence in modern quantum systems, allowing for longer measurement durations and more reliable data acquisition.

Entanglement for Enhanced Sensitivity

Quantum entanglement is a phenomenon in which two or more qubits become interlinked, such that the state of one qubit instantly influences the state of the other, no matter the distance between them. In sensing applications, entanglement can be used to amplify measurement sensitivity and reduce uncertainty. When qubits are entangled, the measurement precision can surpass the classical shot-noise limit and approach the quantum Heisenberg limit. This has practical applications in technologies such as atomic clocks and magnetometers, where entangled particles provide enhanced timekeeping and magnetic field detection capabilities.

Special Quantum States in Sensing

Certain quantum states are particularly useful for increasing measurement precision. One such example is the squeezed state, where the uncertainty in one observable (like position or momentum) is reduced at the expense of increased uncertainty in the conjugate observable. Squeezed states are often used in quantum optics and have already been implemented in instruments like gravitational wave detectors to improve their sensitivity. By engineering such states, scientists can fine-tune sensors to minimize noise in the parameters of interest, resulting in higher-resolution and more accurate measurements.

Platforms for Realizing Qubits

Various physical systems are used to realize qubits, each offering distinct advantages for sensing applications. Trapped ions, superconducting qubits, neutral atoms, and nitrogen-vacancy (NV) centers in diamond are among the most widely utilized platforms. NV centers, for instance, are atomic-scale defects in diamond that exhibit strong sensitivity to magnetic and electric fields, making them ideal for nanoscale imaging and biological sensing. Superconducting qubits are useful in applications that require integration with electronic circuits, while trapped ions offer very long coherence times, making them excellent for timekeeping and gravitational measurements.

Quantum State Readout and Measurement

Measuring quantum states is a crucial step in the sensing process. Unlike classical systems, where measurement simply reveals a property, quantum measurements influence the system by collapsing the wavefunction into one of its basis states. This makes the readout process delicate and probabilistic. To obtain precise results, quantum sensors typically perform repeated measurements and apply statistical methods to interpret outcomes. High-fidelity readout methods such as fluorescence detection, photodetection, or charge readout are employed depending on the qubit platform. Improvements in readout fidelity are vital to improving sensor accuracy and reliability.

Hybrid Quantum Systems in Sensing

Hybrid systems that combine multiple types of qubits or integrate qubits with classical sensing components are being developed to take advantage of the strengths of different technologies. For example, combining solid-state qubits with optical systems enables more robust and scalable sensor designs. Similarly, integrating mechanical resonators with qubits can allow for the detection of extremely weak forces or small displacements. These hybrid systems are paving the way for practical, portable, and multifunctional quantum sensors that can be deployed in a variety of real-world environments, from space exploration to medical diagnostics.

Applications and Future Outlook

The real-world applications of quantum sensors based on qubits and quantum states are vast and rapidly expanding. In medicine, quantum sensors are being explored for their ability to image tissues and cells with unprecedented resolution. In navigation, quantum gyroscopes and accelerometers provide accurate guidance systems that do not rely on GPS. Environmental monitoring also benefits from quantum sensors that can detect subtle changes in magnetic or gravitational fields, offering tools for geological surveys or early-warning systems for natural disasters. As research continues, we can expect even more groundbreaking applications to emerge.

2.2 QUANTUM COHERENCE AND DECOHERENCE

2.2.1 Quantum Coherence

The signature property of quantum matter waves, wave interference, relies on coherence. While initially patterned after optical coherence, the theory and experimental understanding of quantum coherence greatly expanded the topic.

Matter Wave Coherence

The simplest extension of optical coherence applies optical concepts to matter waves. For example, when performing the double-slit experiment with atoms instead of light waves, a sufficiently collimated atomic beam creates a coherent atomic wave-function illuminating both slits. Each slit acts as a separate but in-phase beam contributing to the intensity pattern on a screen. These two contributions give rise to an intensity pattern of bright bands due to constructive interference, interlaced with dark bands due to destructive interference, on a downstream screen. Many variations of this experiment have been demonstrated. As with light, transverse coherence (across the direction of propagation) of matter waves is controlled by collimation. Because light, at all frequencies, travels the same velocity, longitudinal and temporal coherence are linked; in matter waves these are independent. In matter waves, velocity (energy) selection controls longitudinal coherence and pulsing or chopping controls temporal coherence.

Quantum Optics

The discovery of the Hanbury Brown and Twiss effect – correlation of light upon coincidence – triggered Glauber's creation of uniquely quantum coherence analysis. Classical optical coherence becomes a classical limit for first-order quantum coherence; higher degree of coherence leads to many phenomena in quantum optics.

Macroscopic Quantum Coherence

Macroscopic scale quantum coherence leads to novel phenomena, the so-called macroscopic quantum phenomena. For instance, the laser, superconductivity and superfluidity are examples of highly coherent quantum systems whose effects are evident at the macroscopic scale. The macroscopic quantum coherence (off-diagonal long-range order, ODLRO) for superfluidity, and laser light, is related to first-order (1-body) coherence/ODLRO, while superconductivity is related to second-order coherence/ODLRO. (For fermions, such as electrons, only even orders of coherence/ODLRO are possible.) For bosons, a Bose–Einstein condensate is an example of a system exhibiting macroscopic quantum coherence through a multiple occupied single-particle state. The classical electromagnetic field exhibits macroscopic quantum coherence. The most obvious example is the carrier signal for radio and TV. They satisfy Glauber's quantum description of coherence.

Quantum Coherence as a Resource

Recently M. B. Plenio and co-workers constructed an operational formulation of quantum coherence as a resource theory. They introduced coherence monotones analogous to the entanglement monotones. Quantum coherence has been shown to be equivalent to quantum entanglement in the sense that coherence can be faithfully described as entanglement, and conversely that each entanglement measure corresponds to a coherence measure.

2.2.2 Decoherence

Quantum decoherence is the loss of quantum coherence. Quantum decoherence has been studied to understand how quantum systems convert to systems that can be explained by classical mechanics. Beginning out of attempts to extend the understanding of quantum mechanics, the theory has developed in several directions and experimental studies have confirmed some of the key issues. Quantum computing relies on quantum coherence and is one of the primary practical applications of the concept.

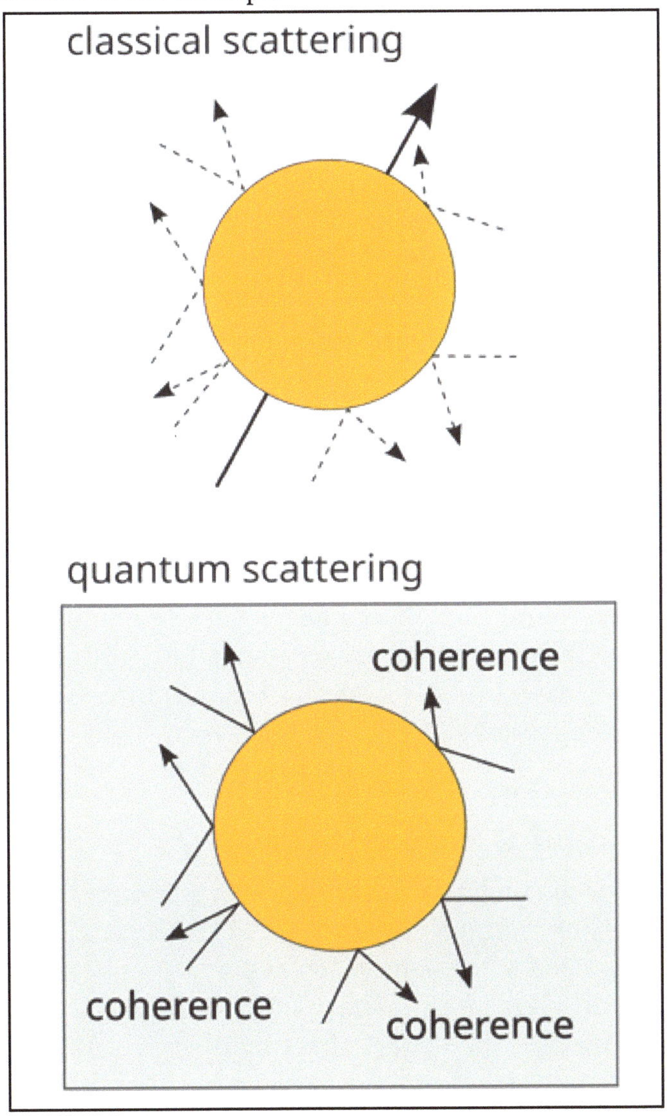

Fig: In Classical Scattering of a Target Body by Environmental Photons, the Motion of the Target Body will not be Changed by the Scattered Photons on the Average

In quantum scattering, the interaction between the scattered photons and the superposed target body will cause them to be entangled, thereby delocalizing the phase coherence from the target body to the whole system, rendering the interference pattern unobservable.

In quantum mechanics, physical systems are described by a mathematical representation called a quantum state. Probabilities for the outcomes of experiments upon a system are calculated by applying the Born rule to the quantum state describing that system. Quantum states are either *pure* or *mixed*; pure states are also known as *wavefunctions*. Assigning a pure state to a quantum system implies certainty about the outcome of some measurement on that system, i.e., that there exists a measurement for which one of the possible outcomes will occur with probability 1. In the absence of outside forces or interactions, a quantum state evolves unitarily over time. Consequently, a pure quantum state remains pure. However, if the system is not perfectly isolated, for example during a measurement, coherence is shared with the environment and appears to be lost with time – a process called quantum decoherence or environmental decoherence. The quantum coherence is not lost but rather mixed with many more degrees of freedom in the environment, analogous to the way energy appears to be lost during friction in classical mechanics when it actually has produced heat in the environment.

Decoherence can be viewed as the loss of information from a system into the environment (often modeled as a heat bath), since every system is loosely coupled with the energetic state of its surroundings. Viewed in isolation, the system's dynamics are non-unitary (although the combined system plus environment evolves in a unitary fashion). Thus the dynamics of the system alone are irreversible. As with any coupling, entanglements are generated between the system and environment. These have the effect of sharing quantum information with—or transferring it to—the surroundings.

History and Interpretation

Relation to Interpretation of Quantum Mechanics

An interpretation of quantum mechanics is an attempt to explain how the mathematical theory of quantum physics might correspond to experienced reality. Decoherence calculations can be done in any interpretation of quantum mechanics, since those calculations are an application of the standard mathematical tools of quantum theory. However, the subject of decoherence has been closely related to the problem of interpretation throughout its history.

Decoherence has been used to understand the possibility of the collapse of the wave function in quantum mechanics. Decoherence does not generate *actual* wave-function collapse. It only provides a framework for *apparent* wave-function collapse, as the components of a quantum system entangle with other quantum systems within the same environment. That is, components of the wave function are decoupled from a coherent system and acquire phases from their immediate surroundings. A total superposition of the global or universal wavefunction still exists (and remains coherent at the global level), but its ultimate fate remains an interpretational issue.

With respect to the measurement problem, decoherence provides an explanation for the transition of the system to a mixture of states that seem to correspond to those states observers perceive. Moreover, observation indicates that this mixture looks like a proper quantum ensemble in a measurement situation, as the measurements lead to the "realization" of precisely one state in the "ensemble".

The philosophical views of Werner Heisenberg and Niels Bohr have often been grouped together as the "Copenhagen interpretation", despite significant divergences between them on important points. In 1955, Heisenberg suggested that the interaction of a system with its surrounding environment would eliminate quantum interference effects. However, Heisenberg did not provide a detailed account of how this might transpire, nor did he make explicit the importance of entanglement in the process.

Origin of the Concepts

Nevill Mott's solution to the iconic Mott problem in 1929 is considered in retrospect to be the first quantum decoherence work. It was cited by the first modern theoretical treatment. Although he did not use the term, the concept of quantum decoherence was first introduced in 1951 by the American physicist David Bohm, who called it the "destruction of interference in the process of measurement". Bohm later used decoherence to handle the measurement process in the de Broglie-Bohm interpretation of quantum theory.

The significance of decoherence was further highlighted in 1970 by the German physicist H. Dieter Zeh, and it has been a subject of active research since the 1980s. Decoherence has been developed into a complete framework, but there is controversy as to whether it solves the measurement problem, as the founders of decoherence theory admit in their seminal papers.

The study of decoherence as a proper subject began in 1970, with H. Dieter Zeh's paper "On the Interpretation of Measurement in Quantum Theory". Zeh regarded the wavefunction as a physical entity, rather than a calculational device or a compendium of statistical information (as is typical for Copenhagen-type interpretations), and he proposed that it should evolve unitarily, in accord with the Schrödinger equation, at all times. Zeh was initially unaware of Hugh Everett III's earlier work, which also proposed a universal wavefunction evolving unitarily; he revised his paper to reference Everett after learning of Everett's "relative-state interpretation" through an article by Bryce DeWitt. (DeWitt was the one who termed Everett's proposal the many-worlds interpretation, by which name it is commonly known.) For Zeh, the question of how to interpret quantum mechanics was of key importance, and an interpretation along the lines of Everett's was the most natural. Partly because of a general disinterest among physicists for interpretational questions, Zeh's work remained comparatively neglected until the early 1980s, when two papers by Wojciech Zurek invigorated the subject. Unlike Zeh's publications, Zurek's articles were fairly agnostic about interpretation, focusing instead on specific problems of density-matrix dynamics. Zurek's interest in decoherence stemmed from furthering Bohr's analysis of the double-slit experiment in his reply to the Einstein–Podolsky–Rosen paradox, work he had undertaken with Bill Wootters, and he has since argued that decoherence brings a kind of rapprochement between Everettian and Copenhagen-type views.

Decoherence does not claim to provide a mechanism for some actual wave-function collapse; rather it puts forth a reasonable framework for the appearance of wave-function collapse. The quantum nature of the system is simply entangled into the environment so that a total superposition of the wave function still exists, but exists—at least for all practical purposes—beyond the realm of measurement. By definition, the claim that a merged but unmeasurable wave function still exists cannot be proven experimentally. Decoherence is needed to understand why a quantum system begins to obey classical probability rules after interacting with its environment (due to the suppression of the interference terms when applying Born's probability rules to the system). Criticism of the adequacy of decoherence theory to solve the measurement problem has been expressed by Anthony Leggett.

Mechanisms

The model requires some familiarity with quantum theory basics. Analogies are made between visualizable classical phase spaces and Hilbert spaces. A more rigorous derivation in Dirac notation shows how decoherence destroys interference effects and the "quantum nature" of systems. Next, the density matrix approach is presented for perspective.

Phase-Space Picture

An *N*-particle system can be represented in non-relativistic quantum mechanics by a wave function $\psi(x_1, x_2, \ldots, x_N)$, where each x_i is a point in 3-dimensional space. This has analogies with the classical phase space. A classical phase space contains a real-valued function in 6*N* dimensions (each particle contributes 3 spatial coordinates and 3 momenta). In this case a "quantum" phase space, on the other hand, involves a complex-valued function on a 3*N*-dimensional space. The position and momenta are represented by operators that do not commute, and ψ lives in the mathematical structure of a Hilbert space. Aside from these differences, however, the rough analogy holds.

Different previously isolated, non-interacting systems occupy different phase spaces. Alternatively we can say that they occupy different lower-dimensional subspaces in the phase space of the joint system. The *effective* dimensionality of a system's phase space is the number of *degrees of freedom* present, which—in non-relativistic models—is 6 times the number of a system's *free* particles. For a macroscopic system this will be a very large dimensionality. When two systems (the environment being one system) start to interact, though, their associated state vectors are no longer constrained to the subspaces. Instead the combined state vector time-evolves a path through the "larger volume", whose dimensionality is the sum of the dimensions of the two subspaces. The extent to which two vectors interfere with each other is a measure of how "close" they are to each other (formally, their overlap or Hilbert space multiplies together) in the phase space. When a system couples to an external environment, the dimensionality of, and hence "volume" available to, the joint state vector increases enormously. Each environmental degree of freedom contributes an extra dimension.

The original system's wave function can be expanded in many different ways as a sum of elements in a quantum superposition. Each expansion corresponds to a projection of the wave vector onto a basis. The basis can be chosen at will. Choosing an expansion where the resulting basis elements interact with the environment in an element-specific way, such elements will—with overwhelming probability—be rapidly separated from each other by their natural unitary time evolution along their own independent paths.

After a very short interaction, there is almost no chance of further interference. The process is effectively irreversible. The different elements effectively become "lost" from each other in the expanded phase space created by coupling with the environment. In phase space, this decoupling is monitored through the Wigner quasi-probability distribution. The original elements are said to have *decohered.* The environment has effectively selected out those expansions or decompositions of the original state vector that decohere (or lose phase coherence) with each other. This is called "environmentally-induced superselection", or einselection. The decohered elements of the system no longer exhibit quantum interference between each other, as in a double-slit experiment. Any elements that decohere from each other via environmental interactions are said to be quantum-entangled with the environment. The converse is not true: not all entangled states are decohered from each other.

Any measuring device or apparatus acts as an environment, since at some stage along the measuring chain, it has to be large enough to be read by humans. It must possess a very large number of hidden degrees of freedom. In effect, the interactions may be considered to be quantum measurements. As a result of an interaction, the wave functions of the system and the measuring device become entangled with each other. Decoherence happens when different portions of the system's wave function become entangled in different ways with the measuring device. For two einselected elements of the entangled system's state to interfere, both the original system and the measuring in both elements device must significantly overlap, in the scalar product sense. If the measuring device has many degrees of freedom, it is *very* unlikely for this to happen.

As a consequence, the system behaves as a classical statistical ensemble of the different elements rather than as a single coherent quantum superposition of them. From the perspective of each ensemble member's measuring device, the system appears to have irreversibly collapsed onto a state with a precise value for the measured attributes, relative to that element. This provides one explanation of how the Born rule coefficients effectively act as probabilities as per the measurement postulate constituting a solution to the quantum measurement problem.

Dirac Notation

Using Dirac notation, let the system initially be in the state

$$|\psi\rangle = \sum_i |i\rangle\langle i|\psi\rangle,$$

The vector basis of the combination of the system and the environment consists of the tensor products of the basis vectors of the two subsystems. Thus, before any interaction between the two subsystems, the joint state can be written as:

$$|\text{before}\rangle = \sum_i |i\rangle|\epsilon\rangle\langle i|\psi\rangle,$$

There are two extremes in the way the system can interact with its environment: either (1) the system loses its distinct identity and merges with the environment (e.g. photons in a cold, dark cavity get converted into molecular excitations within the cavity walls), or (2) the system is not disturbed at all, even though the environment is disturbed (e.g. the idealized non-disturbing measurement). In general, an interaction is a mixture of these two extremes that we examine.

System Absorbed by Environment

If the environment absorbs the system, each element of the total system's basis interacts with the environment such that:

$$|i\rangle|\epsilon\rangle \, evolves \, \text{into} |\epsilon_i\rangle,$$

and so

$$|\text{before}\rangle \, evolves \; into |\text{after}\rangle = \sum_i |\epsilon_i\rangle\langle i|\psi\rangle.$$

The unitarity of time evolution demands that the total state basis remains orthonormal, i.e. the scalar or inner products of the basis vectors must vanish, since $\langle i \mid j \rangle = \delta_{ij}$:

$$\langle \epsilon_i \mid \epsilon_j \rangle = \delta_{ij}.$$

This orthonormality of the environment states is the defining characteristic required for einselection.

System not Disturbed by Environment

In an idealized measurement, the system disturbs the environment, but is itself undisturbed by the environment. In this case, each element of the basis interacts with the environment such that:

$$\mid i \rangle \mid \epsilon \rangle \text{ evolves into the product} \mid i, \epsilon_i \rangle = \mid i \rangle \mid \epsilon_i \rangle,$$

and so

$$\mid \text{before} \rangle \text{ evolves into} \mid \text{after} \rangle = \sum_i \mid i, \epsilon_i \rangle \langle i \mid \psi \rangle.$$

In this case, unitarity demands that:

$$\langle i, \epsilon_i \mid j, \epsilon_j \rangle = \langle i \mid j \rangle \langle \epsilon_i \mid \epsilon_j \rangle = \delta_{ij} \langle \epsilon_i \mid \epsilon_j \rangle = \delta_{ij} \langle \epsilon_i \mid \epsilon_i \rangle = \delta_{ij},$$

As before, this is the defining characteristic for decoherence to become einselection. The approximation becomes more exact as the number of environmental degrees of freedom affected increases.

Note that if the system basis $\mid i \rangle$ were not an einselected basis, then the last condition is trivial, since the disturbed environment is not a function of i ^{i}, and we have the trivial disturbed environment basis $\mid \epsilon_j \rangle = \mid \epsilon' \rangle$. This would correspond to the system basis being degenerate with respect to the environmentally defined measurement observable. For a complex environmental interaction (which would be expected for a typical macroscale interaction) a non-einselected basis would be hard to define.

Loss of Interference and the Transition from Quantum to Classical Probabilities

The utility of decoherence lies in its application to the analysis of probabilities, before and after environmental interaction, and in particular to the vanishing of quantum interference terms after decoherence has occurred. If we ask what is the probability of observing the system making a transition from $\psi \, to \, \phi \, before \, \psi$ has interacted with its environment, then application of the Born probability rule states that the transition probability is the squared modulus of the scalar product of the two states:

$$\text{prob}_{\text{before}}(\psi \to \phi) = \left| \langle \psi \mid \phi \rangle \right|^2 = \left| \sum_i \psi_i^* \phi_i \right|^2 = \sum_i \left| \psi_i^* \phi_i \right|^2 + \sum_{ij; i \neq j} \psi_i^* \psi_j \phi_j^* \phi_i,$$

Non-Unitary Modelling Examples

Decoherence can be modelled as a non-unitary process by which a system couples with its environment (although the combined system plus environment evolves in a unitary fashion). Thus the dynamics of the system alone, treated in isolation, are non-unitary and, as such, are represented by irreversible transformations acting on the system's Hilbert space $\mathcal{H}$. Since the system's dynamics are represented by irreversible representations, then any information present in the quantum system can be lost to the environment or heat bath. Alternatively, the decay of quantum information caused by the coupling of the system to the environment is referred to as decoherence. Thus decoherence is the process by which information of a quantum system is altered by the system's interaction with its environment (which form a closed system), hence creating an entanglement between the system and heat bath (environment). As such, since the system is entangled with its environment in some unknown way, a description of the system by itself cannot be made without also referring to the environment (i.e. without also describing the state of the environment).

Depolarizing

Depolarizing is a non-unitary transformation on a quantum system which maps pure states to mixed states. This is a non-unitary process because any transformation that reverses this process will map states out of their respective Hilbert space thus not preserving positivity (i.e. the original probabilities are mapped to negative probabilities, which is not allowed). The 2-dimensional case of such a transformation would consist of mapping pure states on the surface of the Bloch sphere to mixed states within the Bloch sphere. This would contract the Bloch sphere by some finite amount and the reverse process would expand the Bloch sphere, which cannot happen.

Dissipation

Dissipation is a decohering process by which the populations of quantum states are changed due to entanglement with a bath. An example of this would be a quantum system that can exchange its energy with a bath through the interaction Hamiltonian. If the system is not in its ground state and the bath is at a temperature lower than that of the system's, then the system will give off energy to the bath, and thus higher-energy eigenstates of the system Hamiltonian will decohere to the ground state after cooling and, as such, will all be non-degenerate. Since the states are no longer degenerate, they are not distinguishable, and thus this process is irreversible (non-unitary).

Timescales

Decoherence represents an extremely fast process for macroscopic objects, since these are interacting with many microscopic objects, with an enormous number of degrees of freedom in their natural environment. The process is needed if we are to understand why we tend not to observe quantum behavior in everyday macroscopic objects and why we do see classical fields emerge from the properties of the interaction between matter and radiation for large amounts of matter. The time taken for off-diagonal components of the density matrix to effectively vanish is called the decoherence time. It is typically extremely short for everyday, macroscale processes. A modern basis-independent definition of the decoherence time relies on the short-time behavior of the fidelity between the initial and the time-dependent state or, equivalently, the decay of the purity.

Experimental Observations

Quantitative Measurement

The decoherence rate depends on a number of factors, including temperature or uncertainty in position, and many experiments have tried to measure it depending on the external environment. The process of a quantum superposition gradually obliterated by decoherence was quantitatively measured for the first time by Serge Haroche and his co-workers at the École Normale Supérieure in Paris in 1996. Their approach involved sending individual rubidium atoms, each in a superposition of two states, through a microwave-filled cavity. The two quantum states both cause shifts in the phase of the microwave field, but by different amounts, so that the field itself is also put into a superposition of two states. Due to photon scattering on cavity-mirror imperfection, the cavity field loses phase coherence to the environment. Haroche and his colleagues measured the resulting decoherence via correlations between the states of pairs of atoms sent through the cavity with various time delays between the atoms.

In July 2011, researchers from University of British Columbia and University of California, Santa Barbara showed that applying high magnetic fields to single molecule magnets suppressed two of three known sources of decoherence. They were able to measure the dependence of decoherence on temperature and magnetic field strength.

Prevention

Concept

Decoherence causes the system to lose its quantumness, which invalidates the superposition principle and turns 'quantum' to 'classical'. It is a major challenge in quantum computing.

A real quantum system inevitably meets the surrounding environment, the interaction shows up as noise in physical process. It's extremely sensitive to environmental noise. Electromagnetic fields, temperature fluctuations, and other external perturbations, as well as measurement, lead to decoherence.

Decoherence is a challenge for the practical realization of quantum computers, since such machines are expected to rely heavily on the undisturbed evolution of quantum coherences. They require that the coherence of states be preserved and that decoherence be managed, in order to actually perform quantum computation. Because of decoherence, we need to finish the quantum process before the qubit state is decayed. The physical quantity coherence time is defined as the time that the quantum state holds its superposition principle. The purpose of against decoherence is to extend the coherence time of quantum systems. It will improve the stability of the computing of information.

Methods and Tools

Researchers have developed many methods and tools to mitigate or eliminate the negative influences from decoherence.

Isolation from Environment

The most basic and direct way to reduce decoherence is to prevent the quantum system from interacting with the environment by any type of isolation. Here are some typical examples of isolation methods.

- **High Vacuum:** Placing qubits in an ultra-high vacuum environment to minimize interaction with air molecules.
- **Cryogenic Cooling:** Operating quantum systems at extremely low temperatures to reduce thermal vibrations and noise.
- **Electromagnetic Shielding:** Enclosing quantum systems in materials that block external electromagnetic fields - such as mu-metal or superconducting materials - reduces decoherence caused by unwanted electromagnetic interference.
- **Shielding Cosmic Rays:** In August 2020 scientists reported that ionizing radiation from environmental radioactive materials and cosmic rays may substantially limit the coherence times of qubits if they aren't shielded adequately which may be critical for realizing fault-tolerant superconducting quantum computers in the future.
- **Better Materials:** Fabricating qubits from special materials, like highly pure or isotopically enriched ones, to minimize intrinsic noise of the material, including noise from defects or nuclear spins.
- **Circuit Design:** Optimizing the coherence ability when designing the construction of quantum circuits, similar to the concern in classical circuits.

- **Mechanical and Optical Isolation:** Using equipment like vibration isolation tables and acoustic isolation materials, reducing sources of mechanical noise, and shielding against external light—common in physical experiments.

Quantum Error Correction (QEC)

One of the most powerful tools for combating quantum decoherence is Quantum Error Correction (QEC). QEC schemes encode quantum information redundantly across multiple physical qubits, allowing for the detection and correction of errors without directly measuring the quantum state. These QEC protocols rely on the assumption that errors affect only a small fraction of qubits at any given time, enabling the detection and correction of errors through redundant encoding. Here are some representative QEC protocols.

- **Shor Code:** One of the first quantum error correction codes, it encodes a single qubit into nine physical qubits to protect against both bit-flip and phase-flip errors.
- **Steane Code:** A 7-qubit code that provides error correction for arbitrary errors.
- **Surface Codes:** A more scalable error correction code that uses a 2D lattice of qubits with high threshold for errors.
- **Bosonic Codes:** A type of quantum error-correcting code designed specifically to protect quantum information in continuous-variable systems.

However, QEC comes at a significant cost: it requires a large number of physical qubits to encode a single logical qubit, and fault-tolerant error correction methods introduce additional computational overhead.

Dynamical Decoupling

Dynamical Decoupling (DD) is another typical quantum control technique used against decoherence, especially for systems that are coupled to noisy environments. DD involves applying an external sequence of control pulses to the quantum system at strategically timed intervals to average out environmental interactions. This technique effectively manipulates the irreversible component of quantum systems interact with surrounding environment by the external controllable interactions. Dynamical decoupling has been experimentally demonstrated in various systems, including trapped ions and superconducting qubits. Here are some examples of representative sequences.

- **Spin Echo (SE):** SE is the consisting of a single π-pulse, which inverts the state of system.

- **Periodic Dynamical Decoupling (PDD):** Applying control pulse periodically, PDD averages out the influence of the environment and decoupling the qubit.
- **Carr-Purcell-Meiboom-Gill (CPMG) Sequence:** CPMG is an extension of SE. It applies a series of π-pulses.

2.3 QUANTUM MEASUREMENT AND SENSITIVITY

Quantum measurement and sensitivity are foundational aspects of quantum technologies, particularly in the domain of quantum sensing. Unlike classical measurement techniques that rely on deterministic properties and macroscopic interactions, quantum measurement delves into the realm of probabilities, wavefunctions, and subtle perturbations that can yield extraordinary precision. The concept of quantum sensitivity arises from the ability of quantum systems to detect extremely small changes in physical parameters such as magnetic fields, electric fields, time, gravity, and temperature. This enhanced sensitivity originates from the inherent properties of quantum mechanics, including superposition, entanglement, and coherence, making quantum measurements a powerful tool in both theoretical research and applied science.

Quantum measurement and sensitivity represent a profound leap beyond classical measurement paradigms, driven by the unique features of quantum mechanics. Through phenomena such as superposition, entanglement, and coherence, quantum systems can detect and respond to changes in the environment with extraordinary precision. Although technical challenges remain, ongoing research is rapidly overcoming these obstacles, paving the way for a new era of measurement technology. As quantum sensors become more refined and widespread, they promise to revolutionize our ability to observe and interact with the world at both the microscopic and macroscopic levels.

2.3.1 Understanding Quantum Measurement

In quantum mechanics, the act of measurement is fundamentally different from that in classical physics. When a quantum system is observed, its wavefunction collapses into a definite state. This process is inherently probabilistic, meaning that repeated measurements of identically prepared systems can yield different outcomes, all governed by the system's quantum state prior to measurement. This probabilistic nature of quantum measurement is not a result of experimental error but is intrinsic to the theory itself.

The precision with which a property can be measured is often limited by the Heisenberg uncertainty principle, which states that certain pairs of physical quantities, such as position and momentum or energy and time, cannot both be known with arbitrary precision. Despite these constraints, quantum systems can be harnessed to push the boundaries of what is measurable, especially through cleverly designed protocols and entangled states.

2.3.2 The Role of Superposition in Sensitivity

Superposition is a key quantum phenomenon that enhances measurement sensitivity. A quantum system in a superposition exists simultaneously in multiple states, which allows it to encode information about its environment in a highly nuanced way. When a qubit in superposition interacts with an external field, its state accumulates a phase that reflects the field's characteristics.

This accumulated phase can be read out with extreme precision using interferometric techniques. The sensitivity of the measurement is directly related to how well one can detect changes in this phase. Thus, by preparing and maintaining superposition states, quantum sensors can detect minute differences in external stimuli, often surpassing the limitations of classical systems.

2.3.3 Entanglement and Precision Enhancement

Entanglement, another cornerstone of quantum mechanics, plays a crucial role in increasing measurement precision. Entangled particles share a joint quantum state, such that the measurement of one instantaneously affects the state of the other, regardless of the distance between them. In quantum sensing, entanglement allows for the implementation of quantum-enhanced metrology techniques that achieve higher sensitivity than what is possible with unentangled particles.

For example, using entangled states in an atomic clock or interferometer can reduce the noise associated with independent measurements, enabling detection of extremely small changes in time or phase. This phenomenon is known as quantum noise reduction and is fundamental in achieving sensitivities at or beyond the standard quantum limit.

2.3.4 Heisenberg Limit and Quantum Sensitivity

In classical measurements, the precision of a measurement generally follows the standard quantum limit, which improves with the square root of the number of particles or measurements. However, quantum strategies involving entanglement can push this limit closer to the Heisenberg limit, where precision scales inversely with the number of resources used.

This quadratic improvement represents the ultimate bound imposed by quantum mechanics and defines the theoretical ceiling for sensitivity in quantum measurements. Reaching or approaching the Heisenberg limit is one of the major goals in the design of next-generation quantum sensors, as it would allow the detection of extremely weak forces, tiny frequency shifts, or minimal spatial displacements with minimal error.

2.3.5 Coherence Time and Measurement Accuracy

Quantum coherence, the ability of a quantum system to maintain a well-defined phase relationship over time, is another critical factor that influences sensitivity. Longer coherence times allow quantum sensors to interact with their environment for extended periods without losing their quantum properties. This prolonged interaction leads to better phase accumulation and, consequently, more accurate measurements. However, maintaining coherence is a technical challenge because quantum systems are highly sensitive to noise and decoherence caused by unwanted interactions with the environment. Advanced techniques such as dynamical decoupling, error correction, and environmental shielding are often used to preserve coherence and thereby extend the sensitivity window of quantum devices.

2.3.6 Quantum Measurement Protocols

Several experimental protocols have been developed to extract high-fidelity information from quantum systems. Ramsey interferometry, for instance, is a widely used technique in quantum measurement that involves the application of two coherent pulses to a quantum system separated by a free evolution period. The resulting interference pattern reveals detailed information about phase accumulation and thus about external fields acting on the system. Another approach is quantum non-demolition measurement, which allows repeated observations of a quantum state without significantly disturbing it. These protocols are carefully designed to maximize the amount of information extracted while minimizing decoherence and other sources of measurement error.

2.3.7 Applications of Quantum Measurement and Sensitivity

The high sensitivity of quantum measurement techniques has led to a wide range of applications. Quantum magnetometers can detect extremely weak magnetic fields, useful in brain imaging and geological exploration. Quantum gravimeters can measure gravitational acceleration with high precision, aiding in civil engineering and natural resource mapping. In the medical field, quantum sensors are being explored for use in imaging and diagnostics, offering the potential to detect disease markers at the molecular level. Timekeeping has also benefited immensely, with quantum atomic clocks achieving accuracies that redefine the standard of time measurement, essential for global positioning systems and communication networks.

2.3.8 Challenges in Quantum Measurement

Despite the advantages, several challenges remain in the field of quantum measurement and sensitivity. One major hurdle is the fragility of quantum states, which are susceptible to environmental noise and thermal fluctuations. Scaling up quantum systems to include many qubits or particles while preserving entanglement and coherence is another significant difficulty. Furthermore, the complexity of control and readout mechanisms, as well as the requirement for cryogenic or vacuum environments in some platforms, add layers of technical complexity and cost. Addressing these challenges requires continued innovation in quantum hardware, error correction techniques, and system design.

2.3.9 Future Directions in Quantum Sensitivity

The future of quantum measurement lies in integrating quantum sensors into compact, robust, and deployable systems. Efforts are underway to create portable quantum devices that can operate in real-world environments without the need for extreme isolation. Advances in materials science, photonics, and quantum control are expected to make quantum sensors more accessible and affordable. Additionally, the convergence of quantum sensing with machine learning and data analytics could open new frontiers in real-time adaptive sensing and environmental monitoring. The potential for ultra-sensitive detection capabilities continues to drive research across fields ranging from fundamental physics to industrial automation.

2.4 PRECISION LIMITS: HEISENBERG VS. STANDARD QUANTUM LIMITS

In the realm of quantum measurement, precision is not just a technical challenge but a fundamental concept dictated by the laws of quantum mechanics. The limits to how precisely a physical quantity can be measured in a quantum system are bounded by specific principles. Two of the most important of these are the Standard Quantum Limit (SQL) and the Heisenberg Limit. These two boundaries define the achievable accuracy in measurements that rely on quantum systems, and understanding their implications is essential for the advancement of quantum technologies.

The comparison between these two precision limits illustrates the fundamental constraints and opportunities in quantum sensing, metrology, and information processing. The exploration of quantum precision limits reveals a compelling narrative of theoretical possibility tempered by practical constraints. The Standard Quantum Limit serves as a baseline achievable with unentangled quantum systems, while the Heisenberg Limit represents the aspirational peak enabled by quantum correlations. As technology and understanding advance, the gap between these two bounds continues to narrow. Realizing devices that operate closer to the Heisenberg Limit holds great promise for scientific discovery, medical diagnostics, navigation, and fundamental physics research. However, bridging this gap requires not only improved hardware but also deeper insights into quantum theory, measurement design, and system control. The ongoing dialogue between theoretical physics and experimental engineering will shape the future of quantum precision, leading toward a new era of measurement capabilities.

2.4.1 The Concept of Measurement Precision in Quantum Mechanics

Unlike classical systems, quantum systems are governed by probabilistic laws and wavefunctions that describe the likelihood of finding a particle in a particular state. In classical measurement, precision can be indefinitely improved by refining the tools or increasing the number of observations. However, in quantum systems, there are inherent uncertainties that place a ceiling on the precision with which certain pairs of physical properties—such as position and momentum, or energy and time—can be measured. These uncertainties are not artifacts of technological imperfections but are embedded in the very framework of quantum theory. The endeavor to measure with utmost precision therefore leads directly into the principles of quantum measurement limits, specifically the Standard Quantum Limit and the Heisenberg Limit.

2.4.2 The Standard Quantum Limit: Definition and Origin

The Standard Quantum Limit arises in the context of independent quantum measurements performed on uncorrelated quantum systems. It reflects the statistical noise resulting from repeated measurements on systems that are not entangled or coherently manipulated. According to this limit, the precision of a measurement improves with the square root of the number of independent resources used, such as particles or photons. This is often referred to as shot noise or quantum projection noise. For example, in optical interferometry, if N photons are used to estimate a phase shift, the uncertainty in the phase measurement scales as 1 divided by the square root of N. This statistical bound is common in many quantum experiments and serves as a benchmark for conventional quantum sensing techniques.

2.4.3 The Heisenberg Limit: Pushing the Boundaries of Precision

The Heisenberg Limit represents a more fundamental and stringent boundary that can be approached or reached by exploiting the full potential of quantum mechanics, particularly through the use of entangled states. It describes the ultimate limit of measurement precision, scaling inversely with the number of resources. In the same example of phase estimation, using quantum entanglement or other correlated strategies, the uncertainty in phase can theoretically be reduced to 1 divided by N, surpassing the Standard Quantum Limit. This scaling is called the Heisenberg scaling and is considered the gold standard for quantum-enhanced metrology. The Heisenberg Limit is rooted in the Heisenberg Uncertainty Principle, which places constraints on the simultaneous knowledge of conjugate variables but also enables trade-offs that can be optimized in precision measurements.

2.4.4 The Role of Entanglement in Surpassing the Standard Quantum Limit

Entanglement is a powerful resource that enables the surpassing of the Standard Quantum Limit. In entangled states, the quantum systems exhibit correlations that cannot be explained by classical statistics. These correlations can be harnessed to reduce measurement noise and improve the sensitivity of detection. For example, in atomic clocks or gravitational wave detectors, entangled atoms or photons can be used to enhance phase sensitivity. This improvement comes from the fact that quantum entanglement allows the group of particles to behave as a collective system rather than as independent entities. As a result, the system's response to external perturbations becomes more pronounced, allowing for measurements that approach or even reach the Heisenberg Limit under ideal conditions.

2.4.5 Practical Realizations and Challenges

Despite the theoretical advantages of the Heisenberg Limit, achieving it in practice is a formidable challenge. Creating and maintaining highly entangled states over large systems requires precise control and isolation from environmental noise. Decoherence, technical noise, and imperfections in measurement apparatuses can quickly erode the quantum advantages that entanglement provides. Consequently, most real-world quantum sensors operate close to or slightly better than the Standard Quantum Limit, but far from the Heisenberg Limit. Researchers are actively developing strategies to overcome these limitations through techniques such as quantum error correction, decoherence-free subspaces, and adaptive measurement protocols. These advancements aim to bring experimental performance closer to the ultimate precision bounds predicted by quantum theory.

2.4.6 Implications for Quantum Sensing and Metrology

The distinction between the Standard Quantum Limit and the Heisenberg Limit has profound implications for the design and development of quantum sensors and metrological devices. Systems that operate at or near the Standard Quantum Limit are already used in applications such as atomic clocks, magnetometers, and optical interferometers. However, pushing performance toward the Heisenberg Limit can unlock new capabilities in detecting weak signals, resolving fine structures, and measuring time and frequency with unprecedented accuracy. For instance, next-generation atomic clocks using entangled ions could redefine the standard of timekeeping. Similarly, quantum-enhanced gravitational wave detectors could observe cosmic phenomena with greater sensitivity, opening new windows into astrophysics.

2.4.7 Comparison of SQL and Heisenberg Limit in Different Systems

Different physical systems demonstrate these quantum precision limits in unique ways. In photonic systems, such as interferometers, the SQL is typically encountered when using coherent light sources, while squeezed or entangled light can achieve Heisenberg-limited measurements. In atomic systems, such as atomic ensembles or Bose-Einstein condensates, uncorrelated atoms follow the SQL, whereas spin-squeezed or GHZ (Greenberger–Horne–Zeilinger) states can reach Heisenberg scaling. The specific choice of system, the degree of control over quantum interactions, and the ability to mitigate noise all influence how closely a measurement can approach the Heisenberg Limit. Understanding these dynamics helps researchers select the most appropriate platforms and strategies for high-precision quantum sensing.

2.4.8 Quantum Fisher Information and the Mathematical Framework

To formalize the concept of quantum precision limits, the mathematical tool known as Quantum Fisher Information (QFI) is often employed. QFI quantifies the amount of information that a quantum state carries about an unknown parameter, and it sets an upper bound on the precision of that parameter's estimation. The Cramér-Rao bound, generalized to the quantum domain, states that the variance of any unbiased estimator is bounded below by the inverse of the Quantum Fisher Information. Systems engineered to maximize QFI are thus capable of operating closer to the Heisenberg Limit. This mathematical framework has become a cornerstone in quantum metrology and guides the development of optimal measurement strategies.

2.5 NOISE AND ERROR SOURCES IN QUANTUM SENSING

Quantum sensing has emerged as a revolutionary field that leverages quantum mechanical properties such as superposition, entanglement, and interference to achieve unprecedented sensitivity and precision in detecting physical quantities. These sensors are capable of outperforming classical counterparts in applications such as gravitational wave detection, magnetic field measurement, and timekeeping. However, the quantum systems underlying these sensors are inherently fragile and susceptible to various sources of noise and error.

These disturbances can significantly degrade the performance and accuracy of the sensor. To design reliable and high-performance quantum sensors, it is essential to understand the types of noise that affect them and develop strategies to mitigate their impact. Noise and error are inherent challenges in the practical realization of quantum sensors. While the theoretical potential of quantum sensing is immense, its full realization depends on how effectively these disturbances can be understood, controlled, and mitigated. From intrinsic quantum fluctuations to environmental decoherence and technical imperfections, each type of noise contributes uniquely to the overall performance degradation. Advances in quantum control, material science, and system engineering are gradually pushing the boundaries of what is experimentally achievable. As quantum sensors continue to evolve, their resilience to noise will determine their applicability in real-world scenarios, ranging from medical diagnostics and geological exploration to fundamental physics research. Mastering noise control is therefore not just a technical necessity but a foundational requirement for the next generation of quantum technologies.

2.5.1 Intrinsic Quantum Noise

One of the fundamental sources of noise in quantum sensing arises from the probabilistic nature of quantum mechanics itself. Even in ideal, noise-free environments, quantum systems exhibit inherent uncertainty due to phenomena such as quantum projection noise and shot noise. Quantum projection noise originates when a quantum system in superposition is measured, collapsing into one of the basis states probabilistically. This randomness in outcomes introduces uncertainty into repeated measurements. Similarly, shot noise, often encountered in optical systems, stems from the discrete nature of photons and causes statistical fluctuations in measurement outcomes. These forms of noise set fundamental limits such as the Standard Quantum Limit and are always present, irrespective of environmental conditions.

2.5.2 Decoherence and Environmental Interactions

Decoherence is a dominant source of noise in quantum systems, especially in real-world settings. It occurs when a quantum system interacts with its external environment, causing a loss of coherence between quantum states. The quantum information encoded in superpositions or entanglement rapidly degrades as the system becomes entangled with environmental degrees of freedom, rendering the quantum advantage ineffective.

For example, in atomic sensors, collisions with background gas molecules or fluctuations in magnetic fields can disrupt the coherence of atomic spins. In superconducting systems, thermal noise or electromagnetic interference can introduce energy fluctuations that collapse coherent quantum states. Decoherence thus imposes strict requirements on isolation, shielding, and temperature control in experimental setups.

2.5.3 Technical and Instrumental Noise

Apart from intrinsic and environmental noise, technical noise from instrumentation and control systems also affects quantum sensors. This includes imperfections in laser intensity, frequency drifts, timing jitter, and inaccuracies in waveform generation. These errors can manifest during the preparation, manipulation, or readout of quantum states. For instance, a slight fluctuation in laser frequency used to interrogate atoms in an optical clock can lead to significant phase errors in time measurements. Moreover, any mechanical instability in the experimental apparatus, such as vibrations or thermal expansion, may translate into measurement uncertainty. While these sources are often correctable through improved engineering, they present practical barriers to achieving optimal sensitivity.

2.5.4 Quantum Measurement Back-Action

An important yet subtle source of error in quantum sensing is measurement back-action. According to the principles of quantum mechanics, the act of measurement perturbs the system being observed. In many sensing protocols, particularly those involving weak measurements or repeated observations, the back-action introduces additional noise. For example, in atomic magnetometers or optomechanical sensors, continuous monitoring of quantum observables such as spin or position disturbs the conjugate variable, introducing uncertainty. This back-action sets a trade-off between information gain and system disturbance, which must be optimized to enhance sensor performance without compromising quantum coherence.

2.5.5 Systematic Errors and Calibration Limitations

Systematic errors, unlike random noise, stem from consistent inaccuracies in the experimental design or assumptions. These include biases in measurement apparatus, miscalibrated sensors, or drifts in control parameters over time. In quantum sensing, such errors can be particularly misleading because the measurements often rely on interference patterns or phase shifts that are sensitive to external perturbations. If not properly calibrated, these systematic deviations can be mistaken for genuine signals, leading to incorrect conclusions. Regular calibration against well-characterized reference standards and feedback-based stabilization are necessary to correct these errors and ensure measurement reliability.

2.5.6 Material Imperfections and Fabrication Variability

In solid-state quantum sensors such as nitrogen-vacancy (NV) centers in diamond or superconducting quantum interference devices (SQUIDs), the quality of materials and fabrication processes greatly influences sensor performance. Imperfections at the atomic or microscopic level, such as lattice defects, impurities, or irregularities in layer thickness, can introduce localized noise and energy loss mechanisms. These imperfections create inhomogeneous broadening and reduce coherence times. Furthermore, variations between devices due to fabrication inconsistencies can lead to device-to-device variability, complicating large-scale integration and standardization of quantum sensing platforms. Advanced materials engineering and nanofabrication techniques are therefore critical for minimizing such sources of error.

2.5.7 Thermal Noise and Temperature Fluctuations

Temperature is another crucial factor that affects the stability and coherence of quantum systems. Thermal noise introduces energy into the system that can cause decoherence or unwanted transitions between quantum states. For many quantum sensors, especially those involving superconducting circuits or ultra-cold atoms, operation at cryogenic temperatures is essential to suppress thermal excitations. Even minor temperature fluctuations can introduce phase noise, frequency shifts, or population changes in quantum states. Active temperature stabilization, thermal shielding, and cryogenic techniques are thus commonly employed to maintain the operational integrity of quantum sensors.

2.5.8 Crosstalk and Quantum Control Errors

In multi-qubit or multi-sensor systems, crosstalk can lead to unintended interactions between quantum elements, thereby introducing error. This is particularly relevant in quantum sensor arrays or quantum computing-based sensing applications. Crosstalk can arise from stray electromagnetic fields, shared control lines, or overlapping wavefunctions in closely packed systems. Additionally, errors in quantum gates or control pulses used to manipulate qubits can distort the intended quantum evolution, degrading measurement precision. Robust quantum control protocols, error mitigation techniques, and spatial isolation are necessary to manage such interactions.

2.5.9 Noise Mitigation and Error Correction Strategies

Recognizing and understanding noise sources is the first step toward mitigating their effects. Quantum error correction, originally developed for quantum computing, offers promising methods for protecting quantum information against certain types of noise. By encoding information redundantly across multiple quantum systems and using error-detection algorithms, it is possible to identify and correct errors without collapsing the quantum state. Additionally, techniques such as dynamical decoupling, spin echo, and feedback stabilization can extend coherence times and reduce susceptibility to environmental noise. Engineering improvements, including better vacuum chambers, magnetic shielding, and vibration isolation, further enhance noise resilience.

CHAPTER 3

Types of Quantum Sensors

3.1 QUANTUM METROLOGY AND ITS ROLE IN SENSING

Quantum metrology is the study of making high-resolution and highly sensitive measurements of physical parameters using quantum theory to describe the physical systems, particularly exploiting quantum entanglement and quantum squeezing. This field promises to develop measurement techniques that give better precision than the same measurement performed in a classical framework. Together with quantum hypothesis testing, it represents an important theoretical model at the basis of quantum sensing.

3.1.1 Applications

An important application of particular note is the detection of gravitational radiation in projects such as LIGO or the Virgo interferometer, where high-precision measurements must be made for the relative distance between two widely separated masses.

However, the measurements described by quantum metrology are currently not used in this setting, being difficult to implement. Furthermore, there are other sources of noise affecting the detection of gravitational waves which must be overcome first. Nevertheless, plans may call for the use of quantum metrology in LIGO.

3.1.2 Scaling and the Effect of Noise

A central question of quantum metrology is how the precision, i.e., the variance of the parameter estimation, scales with the number of particles. Classical interferometers cannot overcome the shot-noise limit. This limit is also frequently called standard quantum limit (SQL)

$$(\Delta\theta)^2 \geq \frac{1}{mN},$$

Quantum metrology can reach the Heisenberg limit given by:

$$(\Delta\theta)^2 \geq \frac{1}{mN^2}.$$

However, if uncorrelated local noise is present, then for large particle numbers the scaling of the precision returns to shot-noise scaling $(\Delta\theta)^2 \propto \frac{1}{N}$.

3.1.3 Relation to Quantum Information Science

There are strong links between quantum metrology and quantum information science. It has been shown that quantum entanglement is needed to outperform classical interferometry in magnetometry with a fully polarized ensemble of spins. It has been proved that a similar relation is generally valid for any linear interferometer, independent of the details of the scheme. Moreover, higher and higher levels of multipartite entanglement is needed to achieve a better and better accuracy in parameter estimation. Additionally, entanglement in multiple degrees of freedom of quantum systems (known as "hyperentanglement"), can be used to enhance precision, with enhancement arising from entanglement in each degree of freedom.

3.1.4 Examples

One example of note is the use of the NOON state in a Mach-Zehnder interferometer to perform accurate phase measurements. A similar effect can be produced using less exotic states such as squeezed states. In quantum illumination protocols, two-mode squeezed states are widely studied to overcome the limit of classical states represented in coherent states. In atomic ensembles, spin squeezed states can be used for phase measurements.

3.2 QUANTUM INTERFEROMETERS

3.2.1 Mach-Zehnder Interferometer

The Mach-Zehnder interferometer is a device used to determine the relative phase shift variations between two collimated beams derived by splitting light from a single source. The interferometer has been used, among other things, to measure phase shifts between the two beams caused by a sample or a change in length of one of the paths. The apparatus is named after the physicists Ludwig Mach (the son of Ernst Mach) and Ludwig Zehnder; Zehnder's proposal in an 1891 article was refined by Mach in an 1892 article. Mach-Zehnder interferometry has been demonstrated with electrons as well as with light. The versatility of the Mach-Zehnder configuration has led to its being used in a range of research topics efforts especially in fundamental quantum mechanics.

Design

The Mach-Zehnder interferometer is a highly configurable instrument. In contrast to the well-known Michelson interferometer, each of the well-separated light paths is traversed only once.

If the source has a low coherence length then great care must be taken to equalize the two optical paths. White light in particular requires the optical paths to be simultaneously equalized over all wavelengths, or no fringes will be visible (unless a monochromatic filter is used to isolate a single wavelength). A compensating cell made of the same type of glass as the test cell (so as to have equal optical dispersion) would be placed in the path of the reference beam to match the test cell. Note also the precise orientation of the beam splitters.

The reflecting surfaces of the beam splitters would be oriented so that the test and reference beams pass through an equal amount of glass. In this orientation, the test and reference beams each experience two front-surface reflections, resulting in the same number of phase inversions. The result is that light travels through an equal optical path length in both the test and reference beams leading to constructive interference.

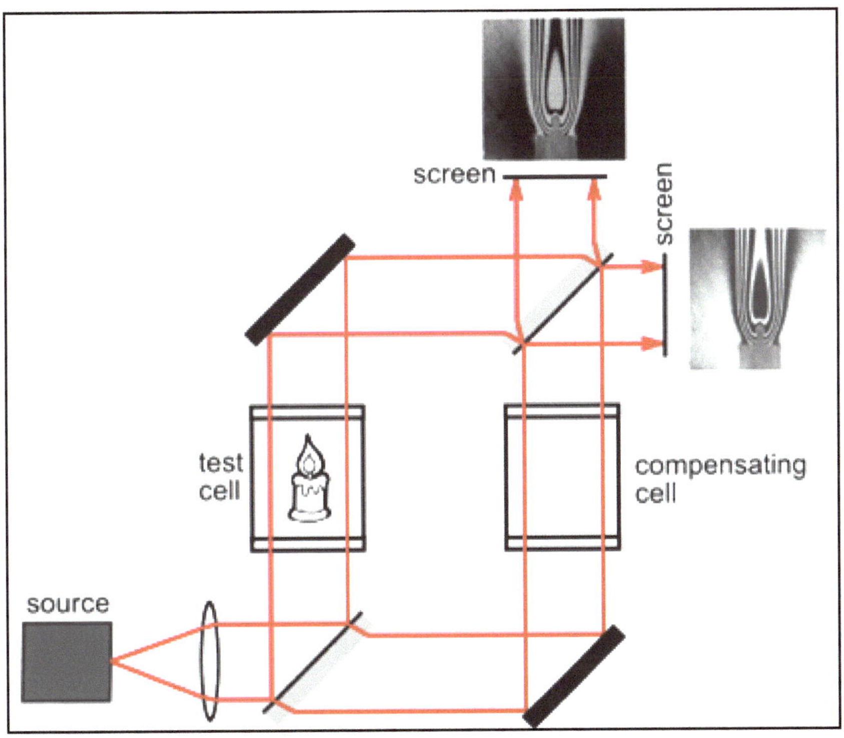

Fig: The Mach–Zehnder Interferometer is Frequently Used in the Fields of Aerodynamics, Plasma Physics and Heat Transfer to Measure Pressure, Density, and Temperature Changes in Gases

In this figure, we imagine analyzing a candle flame. Either output image may be monitored.

Collimated sources result in a nonlocalized fringe pattern. Localized fringes result when an extended source is used. In figure, we see that the fringes can be adjusted so that they are localized in any desired plane. In most cases, the fringes would be adjusted to lie in the same plane as the test object, so that fringes and test object can be photographed together.

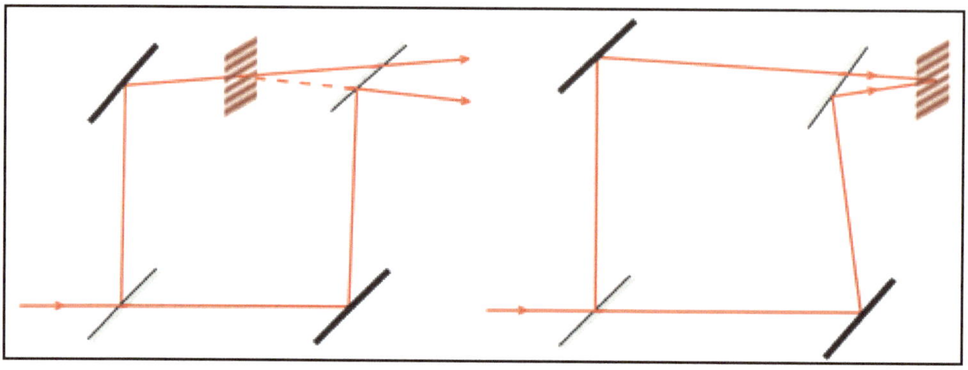

Fig: Localized Fringes Result when an Extended Source is Used in a Mach–Zehnder Interferometer.

Operation

The collimated beam is split by a half-silvered mirror. The two resulting beams (the "sample beam" and the "reference beam") are each reflected by a mirror. The two beams then pass a second half-silvered mirror and enter two detectors. The Fresnel equations for reflection and transmission of a wave at a dielectric imply that there is a phase change for a reflection, when a wave propagating in a lower-refractive index medium reflects from a higher-refractive index medium, but not in the opposite case. A 180° phase shift occurs upon reflection from the front of a mirror, since the medium behind the mirror (glass) has a higher refractive index than the medium the light is traveling in (air). No phase shift accompanies a rear-surface reflection, since the medium behind the mirror (air) has a lower refractive index than the medium the light is traveling in (glass).

The speed of light is lower in media with an index of refraction greater than that of a vacuum, which is 1. Specifically, its speed is: $v = c/n$, where c is the speed of light in vacuum, and n is the index of refraction. This causes a phase shift increase proportional to $(n - 1) \times$ *length traveled*. If k is the constant phase shift incurred by passing through a glass plate on which a mirror resides, a total of $2k$ phase shift occurs when reflecting from the rear of a mirror. This is because light traveling toward the rear of a mirror will enter the glass plate, incurring k phase shift, and then reflect from the mirror with no additional phase shift, since only air is now behind the mirror, and travel again back through the glass plate, incurring an additional k phase shift.

The rule about phase shifts applies to beamsplitters constructed with a dielectric coating and must be modified if a metallic coating is used or when different polarizations are taken into account. Also, in real interferometers, the thicknesses of the beamsplitters may differ, and the path lengths are not necessarily equal. Regardless, in the absence of absorption, conservation of energy guarantees that the two paths must differ by a half-wavelength phase shift. Also beamsplitters that are not 50/50 are frequently employed to improve the interferometer's performance in certain types of measurement.

In figure, in the absence of a sample, both the sample beam (SB) and the reference beam (RB) will arrive in phase at detector 1, yielding constructive interference. Both SB and RB will have undergone a phase shift of (1 × wavelength + k) due to two front-surface reflections and one transmission through a glass plate. At detector 2, in the absence of a sample, the sample beam and reference beam will arrive with a phase difference of half a wavelength, yielding complete destructive interference. The RB arriving at detector 2 will have undergone a phase shift of (0.5 × wavelength + 2k) due to one front-surface reflection and two transmissions. The SB arriving at detector 2 will have undergone a (1 × wavelength + 2k) phase shift due to two front-surface reflections, one rear-surface reflection. Therefore, when there is no sample, only detector 1 receives light. If a sample is placed in the path of the sample beam, the intensities of the beams entering the two detectors will change, allowing the calculation of the phase shift caused by the sample.

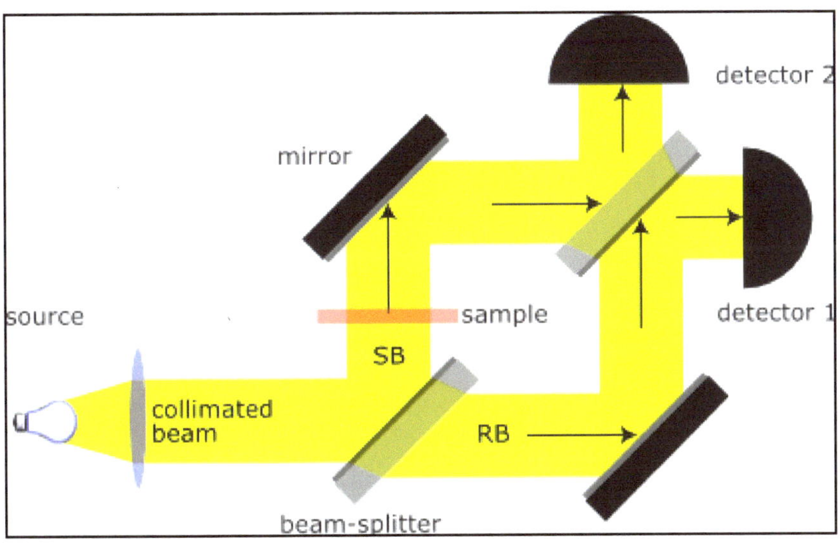

Fig: Effect of a Sample on the Phase of the Output Beams in a Mach–Zehnder Iinterferometer

Uses

The Mach–Zehnder interferometer's relatively large and freely accessible working space, and its flexibility in locating the fringes has made it the interferometer of choice for visualizing flow in wind tunnels and for flow visualization studies in general. It is frequently used in the fields of aerodynamics, plasma physics and heat transfer to measure pressure, density, and temperature changes in gases. Mach–Zehnder interferometers are used in electro-optic modulators, electronic devices used in various fiber-optic communication applications. Mach–Zehnder modulators are incorporated in monolithic integrated circuits and offer well-behaved, high-bandwidth electro-optic amplitude and phase responses over a multiple-gigahertz frequency range. Mach–Zehnder interferometers are also used to study one of the most counterintuitive predictions of quantum mechanics, the phenomenon known as quantum entanglement. The possibility to easily control the features of the light in the reference channel without disturbing the light in the object channel popularized the Mach–Zehnder configuration in holographic interferometry. In particular, optical heterodyne detection with an off-axis, frequency-shifted reference beam ensures good experimental conditions for shot-noise limited holography with video-rate cameras, vibrometry, and laser Doppler imaging of blood flow.

In optical telecommunications it is used as an electro-optic modulator for phase and amplitude modulation of light. Optical computing researchers have proposed using Mach-Zehnder interferometer configurations in optical neural chips for greatly accelerating complex-valued neural network algorithms. The versatility of the Mach–Zehnder configuration has led to its being used in a wide range of fundamental research topics in quantum mechanics, including studies on counterfactual definiteness, quantum entanglement, quantum computation, quantum cryptography, quantum logic, Elitzur–Vaidman bomb tester, the quantum eraser experiment, the quantum Zeno effect, and neutron diffraction.

3.2.2 Sagnac Interferometer

The Sagnac effect, also called Sagnac interference, named after French physicist Georges Sagnac, is a phenomenon encountered in interferometry that is elicited by rotation. The Sagnac effect manifests itself in a setup called a ring interferometer or Sagnac interferometer. A beam of light is split and the two beams are made to follow the same path but in opposite directions. On return to the point of entry the two light beams are allowed to exit the ring and undergo interference. The relative phases of the two exiting beams, and thus the position of the interference fringes, are shifted according to the angular velocity of the apparatus. In other words, when the interferometer is at rest with respect to a nonrotating frame, the light takes the same amount of time to traverse the ring in either direction. However, when the interferometer system is spun, one beam of light has a longer path to travel than the other in order to complete one circuit of the mechanical frame, and so takes longer, resulting in a phase difference between the two beams. Georges Sagnac set up this experiment in 1913 in an attempt to prove the existence of the aether that Einstein's theory of special relativity makes superfluous.

A gimbal mounted mechanical gyroscope remains pointing in the same direction after spinning up, and thus can be used as a rotational reference for an inertial navigation system. With the development of so-called laser gyroscopes and fiber optic gyroscopes based on the Sagnac effect, bulky mechanical gyroscopes can be replaced by those with no moving parts in many modern inertial navigation systems. A conventional gyroscope relies on the principle of conservation of angular momentum whereas the sensitivity of the ring interferometer to rotation arises from the invariance of the speed of light for all inertial frames of reference.

Description and Operation

Typically three or more mirrors are used, so that counter-propagating light beams follow a closed path such as a triangle or square. Alternatively fiber optics can be employed to guide the light through a closed path. If the platform on which the ring interferometer is mounted is rotating, the interference fringes are displaced compared to their position when the platform is not rotating. The amount of displacement is proportional to the angular velocity of the rotating platform. The axis of rotation does not have to be inside the enclosed area.

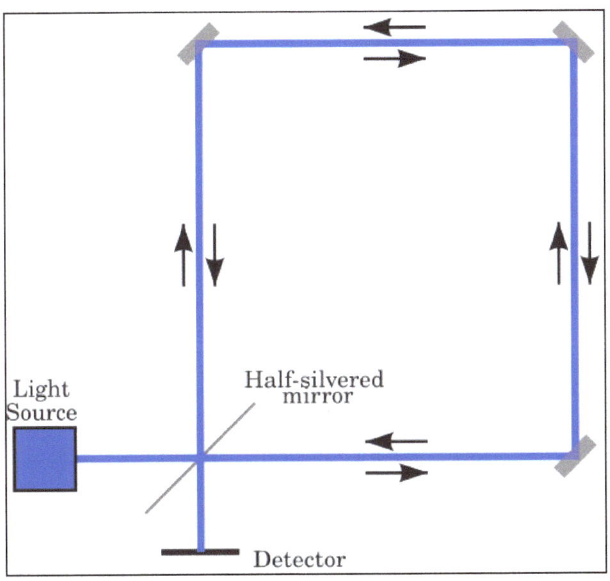

Fig: Schematic Representation of a Sagnac Interferometer

The phase shift of the interference fringes is proportional to the platform's angular frequency ù and is given by a formula originally derived by Sagnac:

$$\Delta\phi \approx \frac{8\pi}{\lambda c} \mathbf{ù} \cdot \mathbf{A}$$

The effect is a consequence of the different times it takes right and left moving light beams to complete a full round trip in the interferometer ring. The difference in travel times, when multiplied by the optical frequency c/λ, determines the phase difference $\Delta\phi$. The rotation thus measured is an absolute rotation, that is, the platform's rotation with respect to an inertial reference frame.

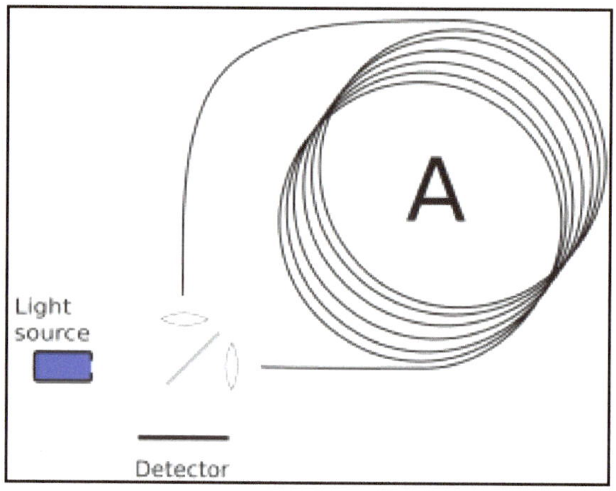

Fig: A Guided Wave Sagnac Interferometer, or Fibre Optic Gyroscope, can be Realized Using an Optical Fiber in a Single or Multiple Loops

History

The Michelson–Morley experiment of 1887 had suggested that the hypothetical luminiferous aether, if it existed, was completely dragged by the Earth. To test this hypothesis, Oliver Lodge in 1897 proposed that a giant ring interferometer be constructed to measure the rotation of the Earth; a similar suggestion was made by Albert Abraham Michelson in 1904. They hoped that with such an interferometer, it would be possible to decide between a stationary aether, versus aethers which are partially or completely dragged by the Earth. That is, if the hypothetical aether were carried along by the Earth (or by the interferometer) the result would be negative, while a stationary aether would give a positive result.

The first description of the Sagnac effect in the framework of special relativity was done by Max von Laue in 1911, two years before Sagnac conducted his experiment. By continuing the theoretical work of Michelson (1904), von Laue confined himself to an inertial frame of reference (which he called a "valid" reference frame), and in a footnote he wrote "a system which rotates in respect to a valid system K^0 is *not* valid". Assuming constant light speed c c, and setting the rotational velocity as ω, he computed the propagation time τ_+ of one ray and τ_- of the counter-propagating ray, and consequently obtained the time difference $\Delta\tau = \tau_+ - \tau_-$. He concluded that this interferometer experiment would indeed produce (when restricted to terms of first order in v/c) the same positive result for both special relativity and the stationary aether (the latter he called "absolute theory" in reference to the 1895-theory of Lorentz). He also concluded that only complete-aether-drag models (such as the ones of Stokes or Hertz) would give a negative result.

The first interferometry experiment aimed at observing the correlation of angular velocity and phase-shift was performed by the French scientist Georges Sagnac in 1913. Its purpose was to detect "the effect of the relative motion of the ether". Sagnac believed that his results constituted proof of the existence of a stationary aether. However, von Laue already showed in 1911 that this effect is consistent with special relativity. Unlike the carefully prepared Michelson–Morley experiment which was set up to prove an aether wind caused by earth drag, the Sagnac experiment could not prove this type of aether wind because a universal aether would affect all parts of the rotating light equally.

Einstein was aware of the phenomenon of the Sagnac effect through the earlier experiments of Franz Harress in 1911. Harress' experiment had been aimed at making measurements of the Fresnel drag of light propagating through moving glass. Not aware of the Sagnac effect, Harress had realized the presence of an "unexpected bias" in his measurements, but was unable to explain its cause. Harress' analysis of the results contained an error, and they were reanalyzed in 1914 by Paul Harzer, who claimed the results were at odds with special relativity. This was rebutted by Einstein. Harress himself died during the First World War, and his results were not publicly available until von Laue persuaded Otto Knopf, whose assistant Harress had been, to publish them in 1920.

Harress' results were published together with an analysis by von Laue, who showed the role of the Sagnac effect in the experiment. Laue said that in the Harress experiment there was a calculable difference in time due to both the dragging of light (which follows from the relativistic velocity addition in moving media, i.e. in moving glass) and "the fact that every part of the rotating apparatus runs away from one ray, while it approaches the other one", i.e. the Sagnac effect. He acknowledged that this latter effect alone could cause the time variance and, therefore, "the accelerations connected with the rotation in no way influence the speed of light".

While Laue's explanation is based on inertial frames, Paul Langevin (1921, 1937) and others described the same effect when viewed from rotating reference frames (in both special and general relativity). So when the Sagnac effect should be described from the viewpoint of a corotating frame, one can use ordinary rotating cylindrical coordinates and apply them to the Minkowski metric, which results into the so-called Born metric or Langevin metric. From these coordinates, one can derive the different arrival times of counter-propagating rays, an effect which was shown by Paul Langevin (1921). Or when these coordinates are used to compute the global speed of light in rotating frames, different apparent light speeds are derived depending on the orientation, an effect which was shown by Langevin in another paper (1937).

Because this apparent variable light speed in rotating frames only arises if rotating coordinates are used, whereas if the Sagnac effect is described from the viewpoint of an external inertial coordinate frame the speed of light of course remains constant – so the Sagnac effect arises no matter whether one uses inertial coordinates or rotating coordinates. That is, special relativity in its original formulation was adapted to inertial coordinate frames, not rotating frames. Albert Einstein in his paper introducing special relativity stated, "light is always propagated in empty space with a definite velocity c which is independent of the state of motion of the emitting body". Einstein specifically stated that light speed is only constant in the vacuum of empty space, using equations that only held in linear and parallel inertial frames. However, when Einstein started to investigate accelerated reference frames, he noticed that "the principle of the constancy of light must be modified" for accelerating frames of reference.

Max von Laue in his 1920 paper gave serious consideration to the effect of General Relativity on the Sagnac effect stating, "General relativity would of course be capable of giving some statements about it, and we want to show at first that no noticeable influences of acceleration are expected according to it." He makes a footnote regarding discussions with German physicist, Wilhelm Wien. The reason for looking at General Relativity is because Einstein's Theory of General Relativity predicted that light would slow down in a gravitational field which is why it could predict the curvature of light around a massive body. Under General Relativity, there is the equivalence principle which states that gravity and acceleration are equivalent. Spinning or accelerating an interferometer creates a gravitational effect. "There are, however, two different types of such [non-inertial] motion; it may for instance be acceleration in a straight line, or circular motion with constant speed." Also, Irwin Shapiro in 1964 explained General Relativity saying, "the speed of a light wave depends on the strength of the gravitational potential along its path". This is called the Shapiro delay. However, since the gravitational field would have to be significant, Laue (1920) concluded it is more likely that the effect is a result of changing the distance of the path by its movement through space. "The beam traveling around the loop in the direction of rotation will have farther to go than the beam traveling counter to the direction of rotation, because during the period of travel the mirrors and detector will all move (slightly) toward the counter-rotating beam and away from the co-rotating beam. Consequently the beams will reach the detector at slightly different times, and slightly out of phase, producing optical interference 'fringes' that can be observed and measured."

In 1926, an ambitious ring interferometry experiment was set up by Albert Michelson and Henry Gale. The aim was to find out whether the rotation of the Earth has an effect on the propagation of light in the vicinity of the Earth. The

Michelson–Gale–Pearson experiment was a very large ring interferometer, (a perimeter of 1.9 kilometer), large enough to detect the angular velocity of the Earth. The outcome of the experiment was that the angular velocity of the Earth as measured by astronomy was confirmed to within measuring accuracy. The ring interferometer of the Michelson–Gale experiment was not calibrated by comparison with an outside reference (which was not possible, because the setup was fixed to the Earth). From its design it could be deduced where the central interference fringe ought to be if there would be zero shift. The measured shift was 230 parts in 1000, with an accuracy of 5 parts in 1000. The predicted shift was 237 parts in 1000.

The Sagnac effect has stimulated a century long debate on its meaning and interpretation, much of this debate being surprising since the effect is perfectly well understood in the context of special relativity.

Theory

Basic Case

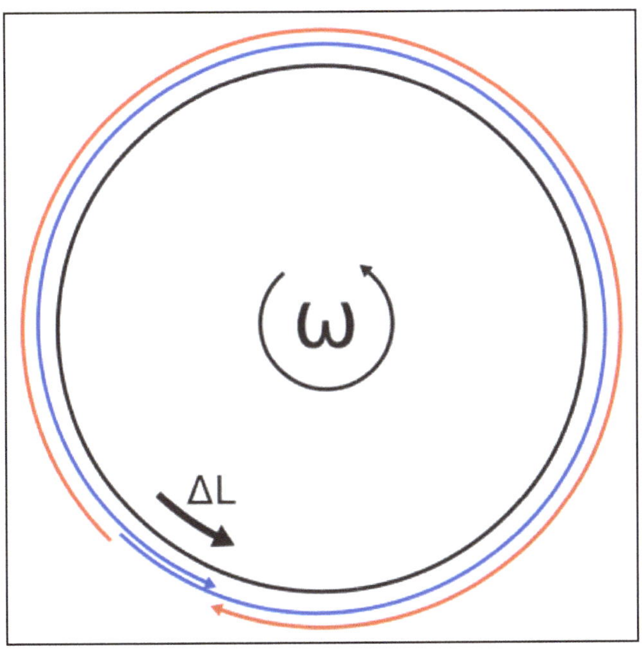

Fig: Light Traveling Opposite Directions Go Different Distances before Reaching the Moving Source

The shift in interference fringes in a ring interferometer can be viewed intuitively as a consequence of the different *distances* that light travels due to the rotation of the ring. The simplest derivation is for a circular ring of radius R, with a refractive index of one, rotating at an angular velocity of ω, but the result is general for loop geometries with other shapes. If a light source emits in both directions from one point on the rotating ring, light traveling in the same direction as the rotation direction needs to travel more than one circumference around the ring before it catches up with the light source from behind.

Although this simple derivation is for a circular ring with an index of refraction of one, the result holds true for any shape of rotating loop with area A. For more complicated shapes, or other refractive index values, the same result can be derived by calculating the optical phase shift in each direction using Fermat's principle and taking into account the different phase velocities for the different propagation directions in an inertial laboratory frame, which can be calculated using relativistic addition of velocities.

We imagine a screen for viewing fringes placed at the light source (or we use a beamsplitter to send light from the source point to the screen). Given a steady light source, interference fringes will form on the screen with a fringe displacement proportional to the time differences required for the two counter-rotating beams to traverse the circuit.

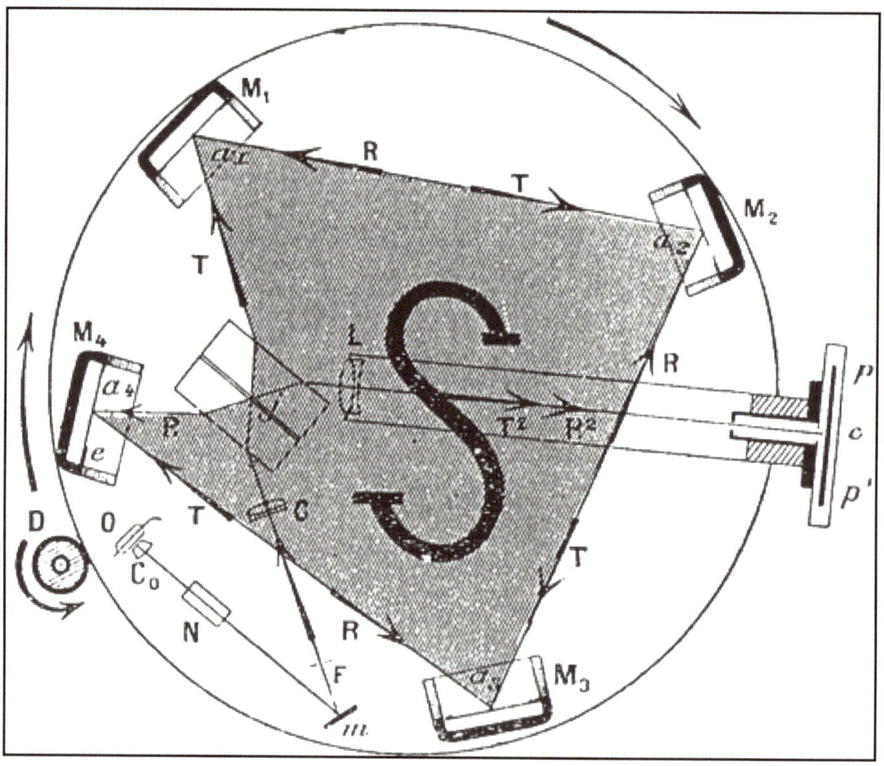

Fig: The Sagnac Area Formula Applies to Any Shape of Loop

At non-relativistic speeds, the Sagnac effect is a simple consequence of the source independence of the speed of light. In other words, the Sagnac experiment does not distinguish between pre-relativistic physics and relativistic physics.

When light propagates in fibre optic cable, the setup is effectively a combination of a Sagnac experiment and the Fizeau experiment. In glass the speed of light is slower than in vacuum, and the optical cable is the moving medium. In that case the relativistic velocity addition rule applies. Pre-relativistic theories of light propagation cannot account for the Fizeau effect. (By 1900 Lorentz could account for the Fizeau effect, but by that time his theory had evolved to a form where in effect it was mathematically equivalent to special relativity.)

Since emitter and detector are traveling at the same speeds, Doppler effects cancel out, so the Sagnac effect does not involve the Doppler effect. In the case of ring laser interferometry, it is important to be aware of this. When the ring laser setup is rotating, the counterpropagating beams undergo frequency shifts in opposite directions. The Sagnac effect is well understood in the context of special relativity where from the rotating light source's point of view the phase difference is due to the line of simultaneity along the light path not forming a closed loop in spacetime.

Generalized Formula

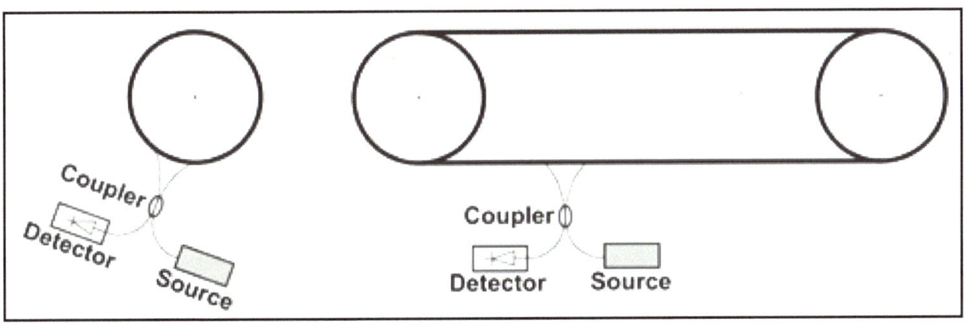

Fig: Conceptually, a Conventional Fibre Optic Gyro (FOG), shown on the Left, can be Divided into Two Semicircular Sections with Extended Fibre Connecting the End Sections as Shown on the Right, Creating a Fibre Optic Conveyor (FOC)

Modified versions of the experiment have been proposed with the light source allowed to move along a (not necessarily circular) light path. This configuration introduces another reason for the phase difference: according to the light source the two signals now follow different paths in space. Some authors refer to this effect as Sagnac effect although in this case the discrepancy need not be due to the lines of simultaneity not forming closed loops.

An example of the modified configuration is shown in figure, the measured phase difference in both a standard fibre optic gyroscope, shown on the left, and a modified fibre optic conveyor, shown on the right, conform to the equation $\Delta t = 2vL/c^2$, whose derivation is based on the constant speed of light. It is evident from this formula that the total time delay is equal to the cumulative time delays along the entire length of fibre, regardless whether the fibre is in a rotating section of the conveyor, or a straight section.

This equation is invalid, however, if the light source's path in space does not follow that of the light signals, for example in the standard rotating platform case

(FOG) but with a non-circular light path. In this case the phase difference formula necessarily involves the area enclosed by the light path due to Stokes' theorem.

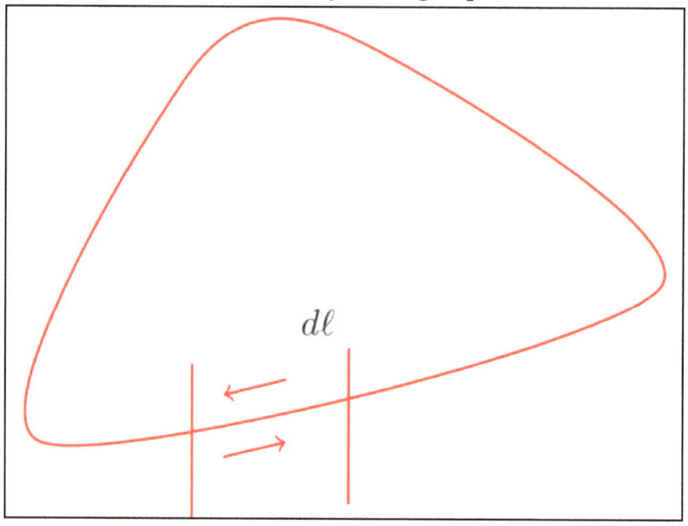

Fig: A Closed Optical Fiber Moving Arbitrarily in Space without Stretching

Consider a ring interferometer where two counter-propagating light beams share a common optical path determined by a loop of an optical fiber. The loop may have an arbitrary shape, and can move arbitrarily in space. The only restriction is that it is not allowed to stretch. (The case of a circular ring interferometer rotating about its center in free space is recovered by taking the index of refraction of the fiber to be 1.)

Applications

A relay of pulses that circumnavigates the Earth, verifying precise synchronization, is also recognized as a case requiring correction for the Sagnac effect. In 1984 a verification was set up that involved three ground stations and several GPS satellites, with relays of signals both going eastward and westward around the world. In the case of a Sagnac interferometer a measure of difference in arrival time is obtained by producing interference fringes, and observing the fringe shift. In the case of a relay of pulses around the world the difference in arrival time is obtained directly from the actual arrival time of the pulses. In both cases the mechanism of the difference in arrival time is the same: the Sagnac effect.

The Hafele–Keating experiment is also recognized as a counterpart to Sagnac effect physics. In the actual Hafele–Keating experiment the mode of transport (long-distance flights) gave rise to time dilation effects of its own, and calculations were needed to separate the various contributions. For the (theoretical) case of clocks that are transported so slowly that time dilation effects arising from the transport are negligible the amount of time difference between the clocks when they arrive back at the starting point will be equal to the time difference that is found for a relay of pulses that travels around the world: 207 nanoseconds.

Practical Uses

The Sagnac effect is employed in current technology. One use is in inertial guidance systems. Ring laser gyroscopes are extremely sensitive to rotations, which need to be accounted for if an inertial guidance system is to return accurate results. The ring laser also can detect the sidereal day, which can also be termed "mode 1".

Global navigation satellite systems (GNSSs) – such as GPS, GLONASS, COMPASS or Galileo – need to take the rotation of the Earth into account in the procedures of using radio signals to synchronize clocks.

Ring Lasers

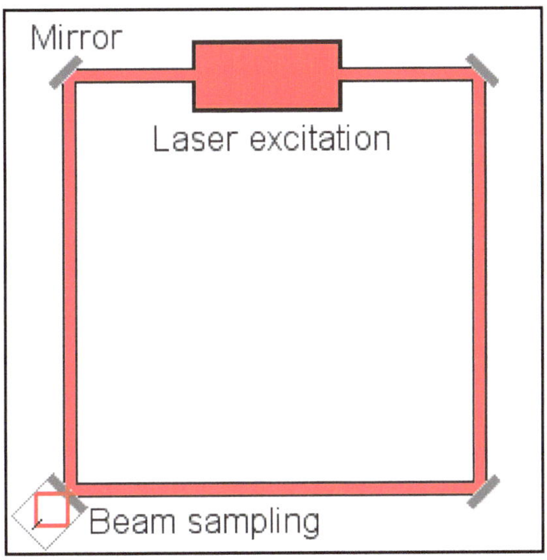

Fig: Schematic Representation of a Ring Laser Setup

Fibre optic gyroscopes are sometimes referred to as 'passive ring interferometers'. A passive ring interferometer uses light entering the setup from outside. The interference pattern that is obtained is a fringe pattern, and what is measured is a phase shift. It is also possible to construct a ring interferometer that is self-contained, based on a completely different arrangement. This is called a ring laser or ring laser gyroscope. The light is generated and sustained by incorporating laser excitation in the path of the light.

To understand what happens in a ring laser cavity, it is helpful to discuss the physics of the laser process in a laser setup with continuous generation of light. As the laser excitation is started, the molecules inside the cavity emit photons, but since the molecules have a thermal velocity, the light inside the laser cavity is at first a range of frequencies, corresponding to the statistical distribution of velocities. The process of stimulated emission makes one frequency quickly outcompete other frequencies, and after that the light is very close to monochromatic.

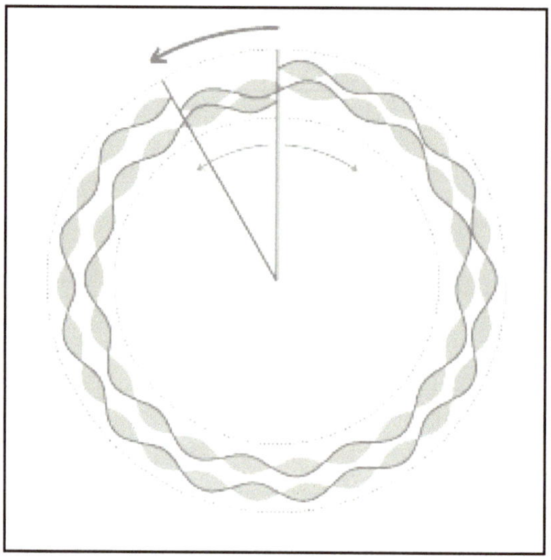

Fig: Schematic Representation of the Frequency Shift when a Ring Laser Interferometer is Rotating. Both the Counterpropagating Light and the Co-Propagating Light go through 12 Cycles of their Frequency

For the sake of simplicity, assume that all emitted photons are emitted in a direction parallel to the ring. Figure illustrates the effect of the ring laser's rotation. In a linear laser, an integer multiple of the wavelength fits the length of the laser cavity. This means that in traveling back and forth the laser light goes through an integer number of *cycles* of its frequency. In the case of a ring laser the same applies: the number of cycles of the laser light's frequency is the same in both directions. This quality of the same number of cycles in both directions is preserved when the ring laser setup is rotating. The image illustrates that there is wavelength shift (hence a frequency shift) in such a way that the number of cycles is the same in both directions of propagation.

By bringing the two frequencies of laser light to interference a beat frequency can be obtained; the beat frequency is the difference between the two frequencies. This beat frequency can be thought of as an interference pattern in time. (The more familiar interference fringes of interferometry are a spatial pattern). The period of this beat frequency is linearly proportional to the angular velocity of the ring laser with respect to inertial space. This is the principle of the ring laser gyroscope, widely used in modern inertial navigation systems.

Zero Point Calibration

In passive ring interferometers, the fringe displacement is proportional to the first derivative of angular position; careful calibration is required to determine the fringe displacement that corresponds to zero angular velocity of the ring interferometer setup. On the other hand, ring laser interferometers do not require calibration to determine the output that corresponds to zero angular velocity. Ring laser interferometers are self-calibrating. The beat frequency will be zero if and only if the ring laser setup is non-rotating with respect to inertial space.

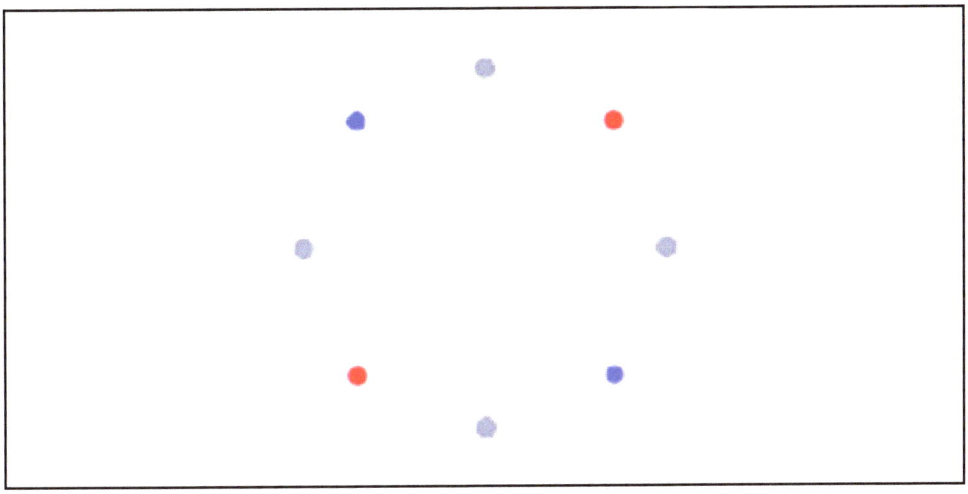

Fig: The Red and Blue Dots Represent Counter-Propagating Photons, the Grey Dots Represent Molecules in the Laser Cavity

Figure illustrates the physical property that makes the ring laser interferometer self-calibrating. The grey dots represent molecules in the laser cavity that act as resonators. Along every section of the ring cavity, the speed of light is the same in both directions. When the ring laser device is rotating, then it rotates with respect to that background. In other words: invariance of the speed of light provides the reference for the self-calibrating property of the ring laser interferometer.

Lock-in

Ring laser gyroscopes suffer from an effect known as "lock-in" at low rotation rates (less than 100°/h). At very low rotation rates, the frequencies of the counter-propagating laser modes become almost identical. In this case, crosstalk between the counter-propagating beams can result in injection locking, so that the standing wave "gets stuck" in a preferred phase, locking the frequency of each beam to each other rather than responding to gradual rotation. By rotationally dithering the laser cavity back and forth through a small angle at a rapid rate (hundreds of hertz), lock-in will only occur during the brief instances where the rotational velocity is close to zero; the errors thereby induced approximately cancel each other between alternating dead periods.

Fibre Optic Gyroscopes versus Ring Laser Gyroscopes

Fibre optic gyros (FOGs) and ring laser gyros (RLGs) both operate by monitoring the difference in propagation time between beams of light traveling in clockwise and counterclockwise directions about a closed optical path. They differ considerably in various cost, reliability, size, weight, power, and other performance characteristics that need to be considered when evaluating these distinct technologies for a particular application.

RLGs require accurate machining, use of precision mirrors, and assembly under clean room conditions. Their mechanical dithering assemblies add somewhat to their weight but not appreciably. RLGs are capable of logging in excess of 100,000 hours of operation in near-room temperature conditions. Their lasers have relatively high power requirements.

Interferometric FOGs are purely solid-state, require no mechanical dithering components, do not require precision machining, have a flexible geometry, and can be made very small. They use many standard components from the telecom industry. In addition, the major optical components of FOGs have proven performance in the telecom industry, with lifespans measured in decades. However, the assembly of multiple optical components into a precision gyro instrument is costly. Analog FOGs offer the lowest possible cost but are limited in performance; digital FOGs offer the wide dynamic ranges and accurate scale factor corrections required for stringent applications. Use of longer and larger coils increases sensitivity at the cost of greater sensitivity to temperature variations and vibrations.

Zero-Area Sagnac Interferometer and Gravitational Wave Detection

The Sagnac topology was actually first described by Michelson in 1886, who employed an even-reflection variant of this interferometer in a repetition of the Fizeau experiment. Michelson noted the extreme stability of the fringes produced by this form of interferometer: White-light fringes were observed immediately upon alignment of the mirrors. In dual-path interferometers, white-light fringes are difficult to obtain since the two path lengths must be matched to within a couple of micrometers (the coherence length of the white light). However, being a common-path interferometer, the Sagnac configuration inherently matches the two path lengths. Likewise Michelson observed that the fringe pattern would remain stable even while holding a lighted match below the optical path; in most interferometers the fringes would shift wildly due to the refractive index fluctuations from the warm air above the match. Sagnac interferometers are almost completely insensitive to displacements of the mirrors or beam-splitter. This characteristic of the Sagnac topology has led to their use in applications requiring exceptionally high stability.

The fringe shift in a Sagnac interferometer due to rotation has a magnitude proportional to the enclosed area of the light path, and this area must be specified in relation to the axis of rotation. Thus the sign of the area of a loop is reversed when the loop is wound in the opposite direction (clockwise or anti-clockwise). A light path that includes loops in both directions, therefore, has a net area given by the difference between the areas of the clockwise and anti-clockwise loops. The special case of two equal but opposite loops is called a *zero-area* Sagnac interferometer. The result is an interferometer that exhibits the stability of the Sagnac topology while being insensitive to rotation.

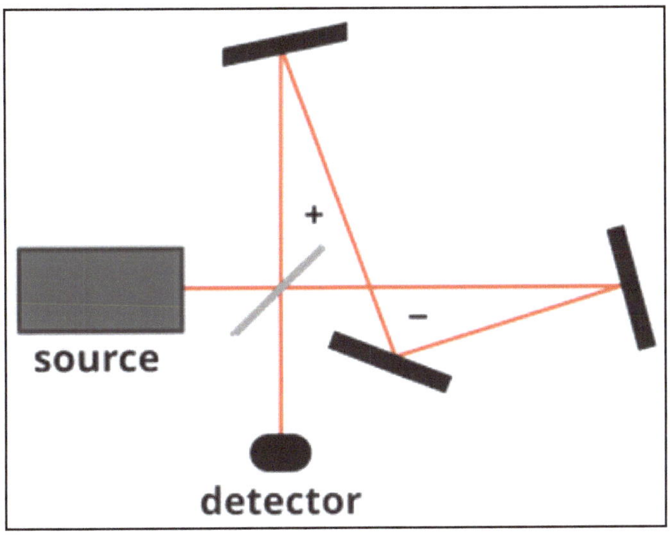

Fig: Zero-Area Sagnac Interferometer

The Laser Interferometer Gravitational-Wave Observatory (LIGO) consisted of two 4-km Michelson–Fabry–Pérot interferometers, and operated at a power level of about 100 watts of laser power at the beam splitter. After an upgrade to Advanced LIGO several kilowatts of laser power are required.

A variety of competing optical systems are being explored for third generation enhancements beyond Advanced LIGO. One of these competing proposals is based on the zero-area Sagnac design. With a light path consisting of two loops of the same area, but in opposite directions, an effective area of zero is obtained thus canceling the Sagnac effect in its usual sense. Although insensitive to low frequency mirror drift, laser frequency variation, reflectivity imbalance between the arms, and thermally induced birefringence, this configuration is nevertheless sensitive to passing gravitational waves at frequencies of astronomical interest. However, many considerations are involved in the choice of an optical system, and despite the zero-area Sagnac's superiority in certain areas, there is as yet no consensus choice of optical system for third generation LIGO.

3.3 QUANTUM MAGNETOMETERS

3.3.1 SQUIDs (Superconducting Quantum Interference Devices)

A SQUID (superconducting quantum interference device) is a very sensitive magnetometer used to measure extremely weak magnetic fields, based on superconducting loops containing Josephson junctions. SQUIDs are sensitive enough to measure fields as low as 5×10^{-18} T with a few days of averaged measurements. Their noise levels are as low as 3 $fT\cdot Hz^{-1/2}$. For comparison, a typical refrigerator magnet produces 0.01 tesla (10^{-2} T), and some processes in animals produce very small magnetic fields between 10^{-9} T and 10^{-6} T. SERF atomic magnetometers, invented in the early 2000s are potentially more sensitive and do not require cryogenic refrigeration but are orders of magnitude larger in size (~1 cm^3) and must be operated in a near-zero magnetic field.

RF SQUID

Fig: A Prototype SQUID

The RF SQUID was invented in 1967 by Robert Jaklevic, John J. Lambe, Arnold Silver, and James Edward Zimmerman at Ford. It is based on the AC Josephson effect and uses only one Josephson junction. It is less sensitive compared to DC SQUID but is cheaper and easier to manufacture in smaller quantities. Most fundamental measurements in biomagnetism, even of extremely small signals, have been made using RF SQUIDS. The RF SQUID is inductively coupled to a resonant tank circuit. Depending on the external magnetic field, as the SQUID operates in the resistive mode, the effective inductance of the tank circuit changes, thus changing the resonant frequency of the tank circuit.

Materials Used

The traditional superconducting materials for SQUIDs are pure niobium or a lead alloy with 10% gold or indium, as pure lead is unstable when its temperature is repeatedly changed. To maintain superconductivity, the entire device needs to operate within a few degrees of absolute zero, cooled with liquid helium.

High-temperature SQUID sensors were developed in the late 1980s. They are made of high-temperature superconductors, particularly YBCO, and are cooled by liquid nitrogen which is cheaper and more easily handled than liquid helium. They are less sensitive than conventional low temperature SQUIDs but good enough for many applications.

In 2006, A proof of concept was shown for CNT-SQUID sensors built with an aluminium loop and a single walled carbon nanotube Josephson junction. The sensors are a few 100 nm in size and operate at 1K or below. Such sensors allow to count spins. In 2022 a SQUID was constructed on magic angle twisted bilayer graphene (MATBG)

Uses

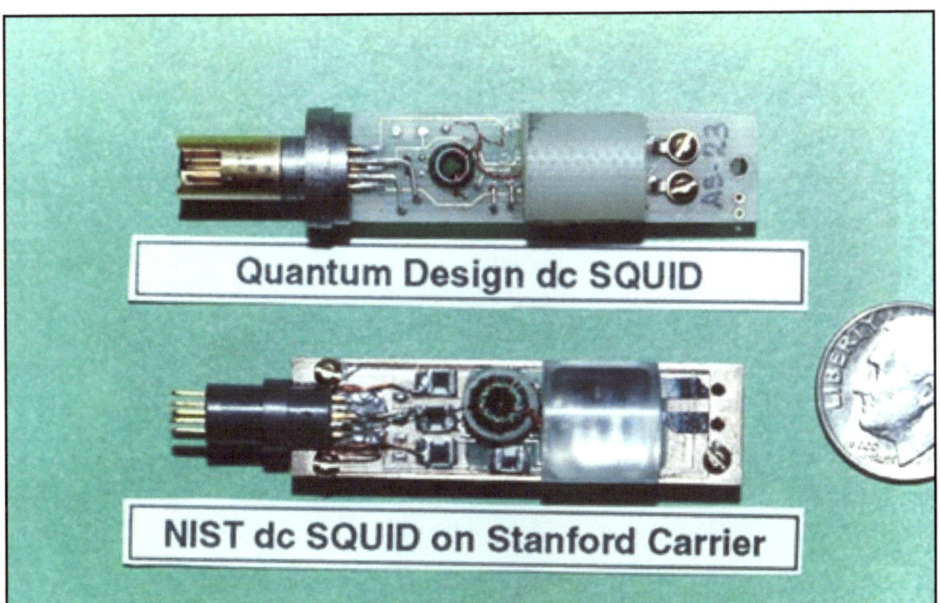

Fig: The Inner Workings of an Early SQUID, Circa 1990

The extreme sensitivity of SQUIDs makes them ideal for studies in biology. Magnetoencephalography (MEG), for example, uses measurements from an array of SQUIDs to make inferences about neural activity inside brains. Because SQUIDs can operate at acquisition rates much higher than the highest temporal frequency of interest in the signals emitted by the brain (kHz), MEG achieves good temporal resolution. Another area where SQUIDs are used is magnetogastrography, which is concerned with recording the weak magnetic fields of the stomach. A novel application of SQUIDs is the magnetic marker monitoring method, which is used to trace the path of orally applied drugs. In the clinical environment SQUIDs are used in cardiology for magnetic field imaging (MFI), which detects the magnetic field of the heart for diagnosis and risk stratification.

Probably the most common commercial use of SQUIDs is in magnetic property measurement systems (MPMS). These are turn-key systems, made by several manufacturers, that measure the magnetic properties of a material sample which typically has a temperature between 300 mK and 400 K. With the decreasing size of SQUID sensors since the last decade, such sensor can equip the tip of an AFM probe. Such device allows simultaneous measurement of roughness of the surface of a sample and the local magnetic flux.

For example, SQUIDs are being used as detectors to perform magnetic resonance imaging (MRI). While high-field MRI uses precession fields of one to several teslas, SQUID-detected MRI uses measurement fields that lie in the microtesla range. In a conventional MRI system, the signal scales as the square of the measurement frequency (and hence precession field): one power of frequency comes from the thermal polarization of the spins at ambient temperature, while the second power of field comes from the fact that the induced voltage in the pickup coil is proportional to the frequency of the precessing magnetization.

In the case of untuned SQUID detection of prepolarized spins, however, the NMR signal strength is independent of precession field, allowing MRI signal detection in extremely weak fields, on the order of Earth's magnetic field. SQUID-detected MRI has advantages over high-field MRI systems, such as the low cost required to build such a system, and its compactness. The principle has been demonstrated by imaging human extremities, and its future application may include tumor screening.

Another application is the scanning SQUID microscope, which uses a SQUID immersed in liquid helium as the probe. The use of SQUIDs in oil prospecting, mineral exploration, earthquake prediction and geothermal energy surveying is becoming more widespread as superconductor technology develops; they are also used as precision movement sensors in a variety of scientific applications, such as the detection of gravitational waves. A SQUID is the sensor in each of the four gyroscopes employed on Gravity Probe B in order to test the limits of the theory of general relativity. A modified RF SQUID was used to observe the dynamical Casimir effect for the first time. SQUIDs constructed from super-cooled niobium wire loops are used as the basis for D-Wave Systems 2000Q quantum computer.

Transition-Edge Sensors

One of the largest uses of SQUIDs is to read out superconducting Transition-edge sensors. Hundreds of thousands of multiplexed SQUIDs coupled to transition-edge sensors are presently being deployed to study the Cosmic microwave background, for X-ray astronomy, to search for dark matter made up of Weakly interacting massive particles, and for spectroscopy at Synchrotron light sources.

Cold Dark Matter

Advanced SQUIDS called near quantum-limited SQUID amplifiers form the basis of the Axion Dark Matter Experiment (ADMX) at the University of Washington. Axions are a prime candidate for cold dark matter.

Proposed Uses

A potential military application exists for use in anti-submarine warfare as a magnetic anomaly detector (MAD) fitted to maritime patrol aircraft. SQUIDs are used in superparamagnetic relaxometry (SPMR), a technology that utilizes the high magnetic field sensitivity of SQUID sensors and the superparamagnetic properties of magnetite nanoparticles. These nanoparticles are paramagnetic; they have no magnetic moment until exposed to an external field where they become ferromagnetic. After removal of the magnetizing field, the nanoparticles decay from a ferromagnetic state to a paramagnetic state, with a time constant that depends upon the particle size and whether they are bound to an external surface. Measurement of the decaying magnetic field by SQUID sensors is used to detect and localize the nanoparticles. Applications for SPMR may include cancer detection.

3.3.2 Nitrogen-Vacancy (NV) Centers in Diamond

The nitrogen-vacancy center (N-V center or NV center) is one of numerous photoluminescent point defects in diamond. Its most explored and useful properties include its spin-dependent photoluminescence (which enables measurement of the electronic spin state using optically detected magnetic resonance), and its relatively long spin coherence at room temperature, lasting up to milliseconds. The NV center energy levels are modified by magnetic fields, electric fields, temperature, and strain, which allow it to serve as a sensor of a variety of physical phenomena. Its atomic size and spin properties can form the basis for useful quantum sensors.

NV centers enable nanoscale measurements of magnetic and electric fields, temperature, and mechanical strain with improved precision. External perturbation sensitivity makes NV centers ideal for applications in biomedicine—such as single-molecule imaging and cellular process modeling. NV centers can also be initialized as qubits and enable the implementation of quantum algorithms and networks. It has also been explored for applications in quantum computing (e.g. for entanglement generation), quantum simulation, and spintronics.

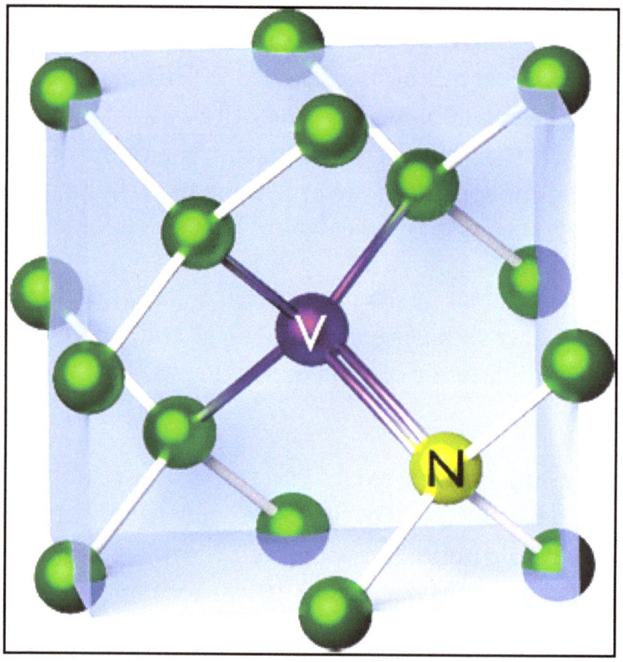

Fig: Simplified Atomic Structure of the NV Center

Structure

The nitrogen-vacancy center is a point defect in the diamond lattice. It consists of a nearest-neighbor pair of a nitrogen atom, which substitutes for a carbon atom, and a lattice vacancy.

Two charge states of this defect, neutral NV^0 and negative NV^-, are known from spectroscopic studies using optical absorption, photoluminescence (PL), electron paramagnetic resonance (EPR) and optically detected magnetic resonance (ODMR), which can be viewed as a hybrid of PL and EPR; most details of the structure originate from EPR. The nitrogen atom on one hand has five valence electrons. Three of them are covalently bonded to the carbon atoms, while the other two remain non-bonded and are called a lone pair. The vacancy on the other hand has three unpaired electrons. Two of them form a quasi covalent bond and one remains unpaired. The overall symmetry, however, is axial (trigonal C_{3V}); one can visualize this by imagining the three unpaired vacancy electrons continuously exchanging their roles.

The NV^0 thus has one unpaired electron and is paramagnetic. However, despite extensive efforts, electron paramagnetic resonance signals from NV^0 avoided detection for decades until 2008. Optical excitation is required to bring the NV^0 defect into the EPR-detectable excited state; the signals from the ground state are presumably too broad for EPR detection.

The NV^0 centers can be converted into NV^- by changing the Fermi level position. This can be achieved by applying external voltage to a p-n junction made from doped diamond, e.g., in a Schottky diode.

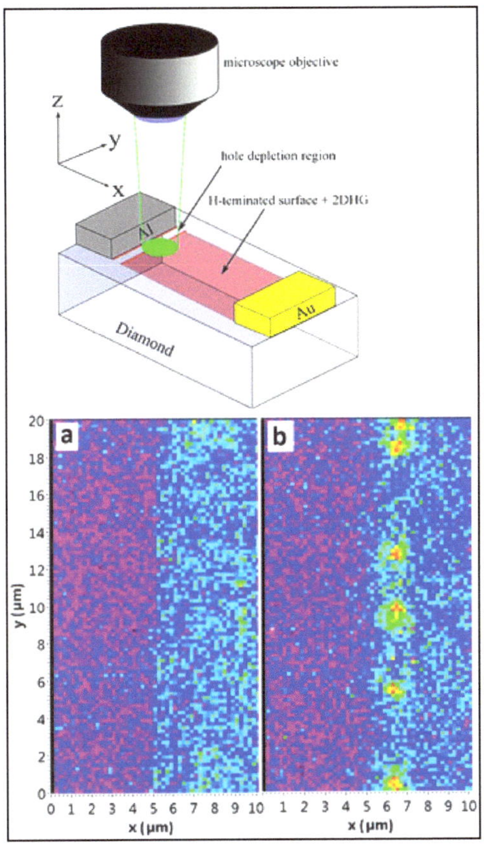

Fig: Bottom Images are Spatial Photoluminescence (PL) Maps before and after Application of +20 V Voltage to a Planar Schottky Diode and the Top Image Outlines the Experiment

In the negative charge state NV^-, an extra electron is located at the vacancy site forming a spin S=1 pair with one of the vacancy electrons. This extra electron induces spin triplet ground states of the form $|^3A\rangle$ and excited states of the form $|^3E\rangle$. There is an additional metastable state that exists between these spin triplets, that often manifests as a singlet. These states play a crucial role in enabling ground state depletion (GSD) microscopy. As in NV^0, the vacancy electrons are "exchanging roles" preserving the overall trigonal symmetry. This NV^- state is what is commonly, and somewhat incorrectly, called "the nitrogen-vacancy center". The neutral state is not generally used for quantum technology.

The NV centers are randomly oriented within a diamond crystal. Ion implantation techniques can enable their artificial creation in predetermined positions.

Production

Nitrogen-vacancy centers are typically produced from single substitutional nitrogen centers (called C or P1 centers in diamond literature) by irradiation followed by annealing at temperatures above 700 °C. A wide range of high-energy particles is suitable for such irradiation, including electrons, protons, neutrons, ions, and gamma photons. Irradiation produces lattice vacancies, which are a part of NV centers. Those vacancies are immobile at room temperature, and annealing is required to move them. Single substitutional nitrogen produces strain in the diamond lattice; it therefore efficiently captures moving vacancies, producing the NV centers.

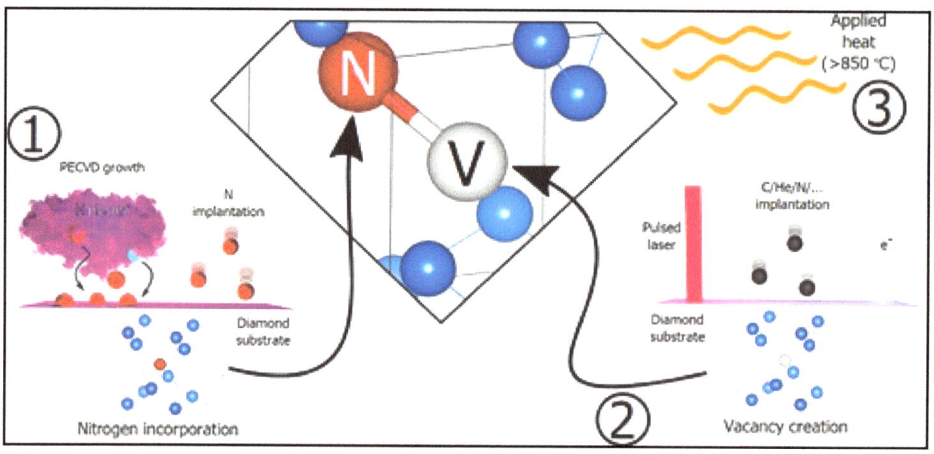

Fig: Production of Nitrogen-Vacancy Centers in Diamond may Require Several Steps

First, nitrogen must be introduced into the diamond lattice which can be accomplished via ion implantation or CVD delta doping. Secondly, vacancies must be introduced, which can be accomplished via laser irradiation, ion implantation, or electron irradiation. Alternatively, during the nitrogen introduction step, vacancies may also be introduced. Finally, a high-temperature annealing step can help promote NV formation.

During chemical vapor deposition of diamond, a small fraction of single substitutional nitrogen impurity (typically <0.5%) traps vacancies generated as a result of the plasma synthesis. Such nitrogen-vacancy centers are preferentially aligned to the growth direction. Delta doping of nitrogen during CVD growth can be used to create two-dimensional ensembles of NV centers near the diamond surface for enhanced sensing or simulation.

Diamond is notorious for having a relatively large lattice strain. Strain splits and shifts optical transitions from individual centers resulting in broad lines in the ensembles of centers. Special care is taken to produce extremely sharp NV lines (line width ~10 MHz) required for most experiments: high-quality, natural (type IIa) diamonds are selected, although synthetic diamonds are preferential. Many of them already have sufficient concentrations of grown-in NV centers and are suitable for applications. If not, they are irradiated by high-energy particles and annealed. Selection of a certain irradiation dose allows tuning the concentration of produced NV centers such that individual NV centers are separated by micrometre-large distances. Then, individual NV centers can be studied with standard optical microscopes or, better, near-field scanning optical microscopes having sub-micrometre resolution.

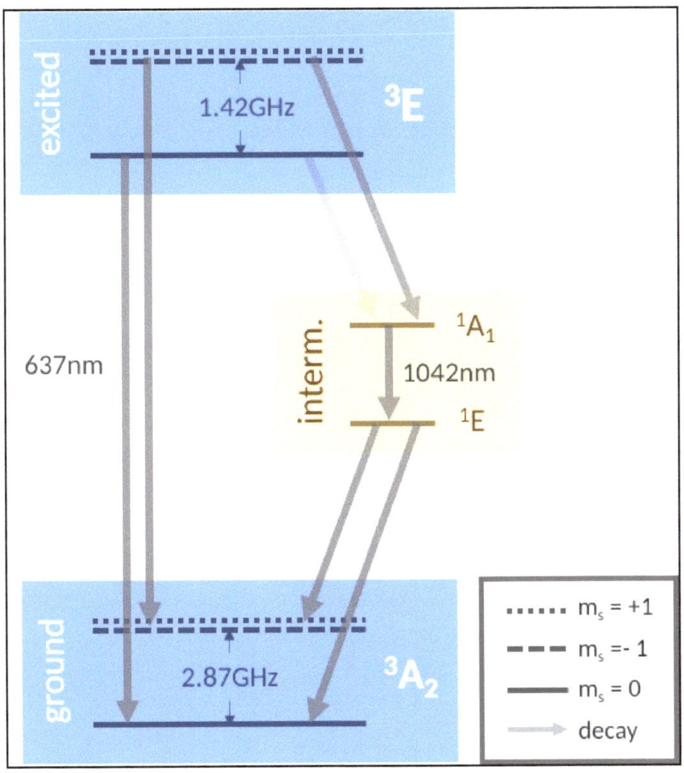

Fig: Schematic Energy Level Structure of the NV Center

Electron transitions between the ground ^{3}A and excited ^{3}E states, separated by 1.945 eV (637 nm), produce absorption and luminescence. The ^{3}A state is split by 2.87 GHz and the ^{3}E state by 1.42 GHz. Numbers 0, ±1 indicate spin quantum number m_s; splitting due to the orbital degeneracy is not shown.

Energy Level Structure

The NV center has a ground-state triplet (^{3}A), an excited-state triplet (^{3}E) and two intermediate-state singlets (^{1}A and ^{1}E). Both ^{3}A and ^{3}E contain $m_s = \pm1$ spin states, in which the two electron spins are aligned (either up, such that $m_s = +1$ or down, such that $m_s = -1$), and an $m_s = 0$ spin state where the electron spins are antiparallel. Due to the magnetic interaction, the energy of the $m_s = \pm1$ states is higher than that of the $m_s = 0$ state. ^{1}A and ^{1}E only contain a spin state singlet each with $m_s = 0$.

If an external magnetic field is applied along the defect axis (the axis which aligns with the nitrogen atom and the vacancy) of the NV center, it does not affect the $m_s = 0$ states, but it splits the $m_s = \pm1$ levels (Zeeman effect). Similarly the following other properties of the environment influence the energy level diagram (further discussed under #Effects of external fields):

- Amplitude and orientation of a static magnetic field splits the $m_s = \pm1$ levels in the ground and excited states.
- Amplitude and orientation of elastic (strain) or electric fields have a much smaller but also more complex effects on the different levels.
- Continuous-wave microwave radiation (applied in resonance with the transition between $m_s = 0$ and (one of the) $m_s = \pm1$ states) changes the population of the sublevels within the ground and excited state.
- A tunable laser can selectively excite certain sublevels of the ground and excited states.
- Surrounding spins and spin–orbit interaction will modulate the magnetic field experienced by the NV center.
- Temperature and pressure affect different parts of the spectrum including the shift between ground and excited states.

It was not this structure alone, but a combination of several favorable factors (previous knowledge, easy production, biocompatibility, simple initialisation, use at room temperature etc.) which suggested the use of the NV center as a qubit and quantum sensor.

Optical Properties

NV centers emit bright red light (^{3}E→^{3}A transitions), if excited off-resonantly by visible green light (^{3}A →^{3}E transitions). This can be done with convenient light sources such as argon or krypton lasers, frequency doubled Nd:YAG lasers, dye lasers, or He-Ne lasers.

As the relaxation time from the excited state is small (~10 ns), the emission happens almost instantly after the excitation. At room temperature the NV center's optical spectrum exhibits no sharp peaks due to thermal broadening. However, cooling the NV centers with liquid nitrogen or liquid helium dramatically narrows the lines down to a width of a few MHz. At low temperature it also becomes possible to specifically address the zero-phonon line (ZPL). An important property of the luminescence from individual NV centers is its high temporal stability. Whereas many single-molecular emitters bleach (i.e. change their charge state and become dark) after emission of 10^6–10^8 photons, bleaching is unlikely for NV centers at room temperature. Strong laser illumination, however, may also convert some NV^- into NV^0 centers.

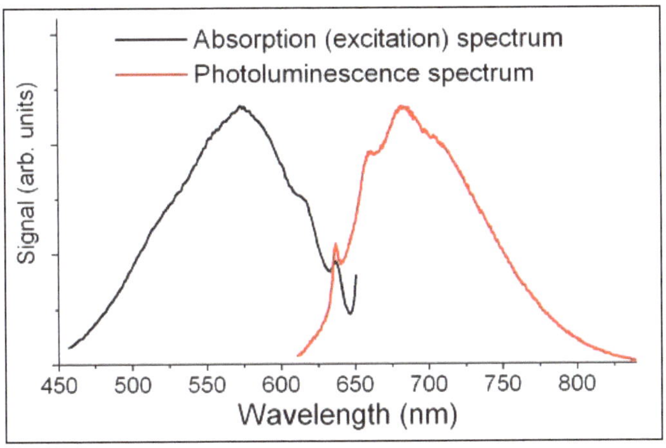

Fig: Optical Absorption and Emission of the NV– Center at Room Temperature

Because of these properties, the ideal technique to address the NV centers is confocal microscopy, both at room temperature and at low temperature.

State Manipulation

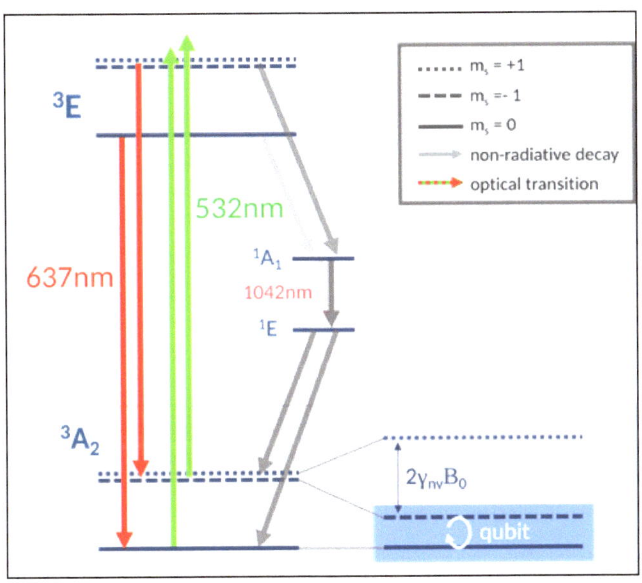

Fig: Spin Dynamics in the NV Center in Diamond

The primary transition between the ground and excited state triplets is spin conserving. Decay via the intermediate singlets gives rise to spin polarization by converting spin from $m_s = \pm 1$ to $m_s = 0$. Both absorption and emission wavelengths are indicated, since they differ due to Stokes shift. Furthermore, the effect of a static magnetic field B_0 along the defect axis and the resulting Zeeman shift is indicated. Here, γ_{nv} refers to the gyromagnetic ratio of the NV center. In many applications two of the ground-state levels are then used as a qubit. Transitions in this effective two-level system, may be induced using a microwave field. 3E-1A and 1E-3A are non-radiative transitions.

Optical Spin Manipulation

Optical transitions must preserve the total spin and occur only between levels of the same total spin. Specifically, transitions between the ground and excited states (with equal spin) can be induced using a green laser with a wavelength of 546 nm. Transitions $^3E \rightarrow {}^1A$ and $^1E \rightarrow {}^3A$ are non-radiative, while $^1A \rightarrow {}^1E$ has both a non-radiative and infrared decay path. The diagram on the right shows the multi-electronic states of the NV center labeled according to their symmetry (E or A) and their spin state (3 for a triplet (S=1) and 1 for a singlet (S=0)). There are two triplet states and two intermediate singlet states.

Spin-State Initialisation:

- An important property of the non-radiative transition between ^3E and ^1A is that it is stronger for $m_s = \pm 1$ and weaker for $m_s = 0$. This provides the basis a very useful manipulation strategy, which is called spin state initialisation (or optical spin-polarization). To understand the process, first consider an off-resonance excitation which has a higher frequency (typically 2.32 eV (532 nm)) than the frequencies of all transitions and thus lies in the vibronic bands for all transitions. By using a pulse of this wavelength, one can excite all spin states from ^3A to ^3E. An NV center in the ground state with $m_s = 0$ will be excited to the corresponding excited state with $m_s = 0$ due to the conservation of spin. Afterwards it decays back to its original state. For a ground state with $m_s = \pm 1$, the situation is different. After the excitation, it has a relatively high probability to decay into the intermediate state ^1A by non-radiative transition and further into the ground state with $m_s = 0$. After many cycles, the state of the NV center (independently of whether it started in $m_s = 0$ or $m_s = \pm 1$) will end up in the $m_s = 0$ ground state. This process can be used to initialize the quantum state of a qubit for quantum information processing or quantum sensing.
- Sometimes the polarisability of the NV center is explained by the claim that the transition from ^1E to the ground state with $m_s = \pm 1$ is small,

compared to the transition to $m_s = 0$. However, it has been shown that the comparatively low decay probability for $m_s = 0$ states w.r.t. $m_s = \pm1$ states into ^1A is enough to explain the polarization.

Effects of External Fields

Microwave Spin Manipulation:

- The energy difference between the $m_s = 0$ and $m_s = 1\pm$ states corresponds to the microwave regime. Population can be transferred between the states by applying a resonant magnetic field perpendicular to the defect axis. Numerous dynamic effects (spin echo, Rabi oscillations, etc.) can be exploited by applying a carefully designed sequence of microwave pulses. Such protocols are rather important for the practical realization of quantum computers. By manipulating the population, it is possible to shift the NV center into a more sensitive or stable state. Its own resulting fluctuating fields may also be used to influence the surrounding nuclei or protect the NV center itself from noise. This is typically done using a wire loop (microwave antenna) which creates an oscillating magnetic field.

Optical Manipulation:

- There are inherent difficulties in achieving miniaturization and effective error reduction in microwave and radio frequency driven spin manipulation techniques. This poses special challenge on application of spin based quantum sensors on sensing electric and magnetic field or any physical phenomena at nanoscale level. The recent developments in microwave-free and optically driven methods pave the way towards energy efficient and coherent quantum sensing. This technique is based on coherent mapping of the spin states of the Nitrogen nucleus to that of the NV center under the application of external magnetic field transverse to the NV symmetry axis. The optical pumping then prepares the system in a coherent superposition state which is the key element in a quantum network.

Influence of External Factors:

- If a magnetic field is oriented along the defect axis it leads to Zeeman splitting separating the $m_s = +1$ from the $m_s = -1$ states. This technique is used to lift the degeneracy and use only two of the spin states (usually the ground states with $m_s = -1$ and $m_s = 0$) as a qubit. Population can then be transferred between them using a microwave field. In the specific instance that the magnetic field reaches 1027 G (or 508 G) then the $m_s = -1$ and $m_s = 0$ states in the ground (or excited) state become equal in energy (Ground/Excited State Level Anticrossing). The following strong interaction results in so-called spin polarization, which strongly affects the intensity of optical absorption and luminescence transitions involving those states.

- Importantly, this splitting can be modulated by applying an external electric field though the physics of the splitting is somewhat more complex. Nevertheless, an important practical outcome is that the intensity and position of the luminescence lines is modulated. Strain has a similar effect on the NV center as electric fields.
- There is an additional splitting of the m_s = 1± energy levels, which originates from the hyperfine interaction between surrounding nuclear spins and the NV center. These nuclear spins create magnetic and electric fields of their own leading to further distortions of the NV spectrum. Also the NV center's own spin–orbit interaction and orbital degeneracy leads to additional level splitting in the excited ^3E state.
- Temperature and pressure directly influence the zero-field term of the NV center leading to a shift between the ground and excited state levels.

Charge State Manipulation

It is also possible to switch the charge state of the NV center (i.e. between NV^-, NV^+ and NV^0) by applying a gate voltage. The gate voltage electrically shifts the Fermi level at the diamond surface and changes its surface band bending. Upon varying the gate voltage, individual centers are allowed to switch from an unknown non-fluorescent state to the neutral charge state NV^0. The ensemble of centers can be transitioned from NV^0 to the qubit state NV^-. The diamond surface termination additionally influences the charge state of near-surface NV centers. Oxygen termination is known to stabilize the NV^-state by reducing surface conductivity and mitigating band bending This improves charge state stability and coherence. In a similar capacity, nitrogen termination also affects surface properties and can optimize NV centers for specific sensing applications.

Optical excitation methods additionally play a role in charge state manipulation. Illumination with specific wavelengths can induce transitions between charge states. Near-infrared light at 1064 nm has been shown to convert NV^0 to NV^-, enhancing photoluminescence. Additionally, it has been demonstrated that NV^+ centers can be switched to NV^0 by photons with energies ≥ 1.23 eV.

Applications

- Schematics of experimental setup. The bottom lens excites a diamond nanocrystal with a green laser light and collects photoluminescence (PL). The crystal hosts an NV center and is attached to the AFM tip.

A wire on the sample surface serves as the microwave source (mw). The temperature of the cantilever T_h is determined from the applied current and voltage.

- ODMR spectra of the NV center at three temperatures. The line splitting originates from a ~1 mT applied magnetic field.
- Thermal conductivity image of a gold letter E on sapphire. White circles indicate features that do not correlate with the AFM topography.
- PL image of the AFM cantilever end and tip where the diamond nanocrystal appears as the bright spot. (e) Zoomed PL image of the NV center in d.

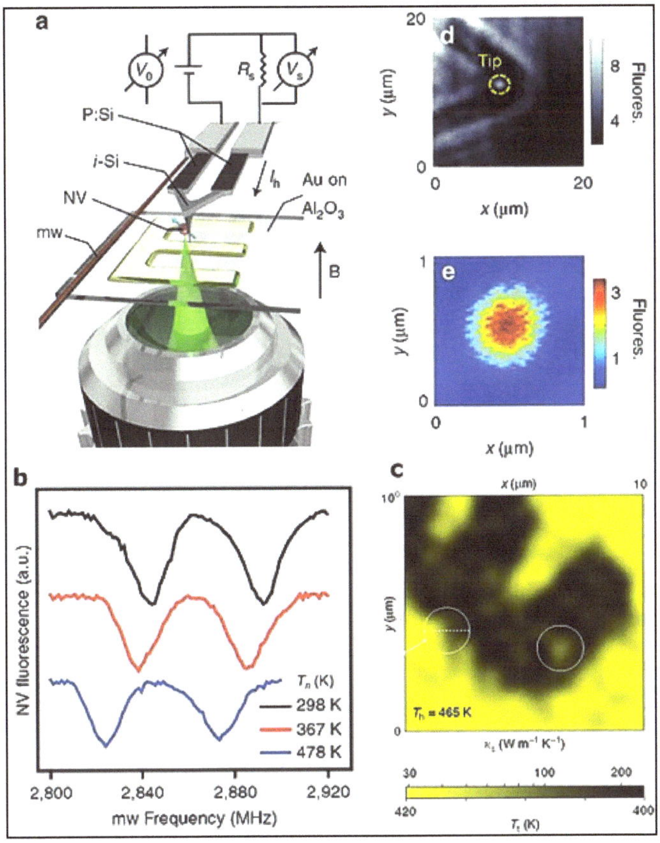

Fig: Scanning Thermal Microscopy using the NV Center

The spectral shape and intensity of the optical signals from the NV⁻ centers are sensitive to external perturbation, such as temperature, strain, electric and magnetic field. However, the use of spectral shape for sensing those perturbation is impractical, as the diamond would have to be cooled to cryogenic temperatures to sharpen the NV⁻ signals. A more realistic approach is to use luminescence intensity (rather than lineshape), which exhibits a sharp resonance when a microwave frequency is applied to diamond that matches the splitting of the ground-state levels. The resulting optically detected magnetic resonance signals are sharp even at room temperature, and can be used in miniature sensors. Such sensors can detect magnetic fields of a few nanotesla or electric fields of about 10 V/cm at kilohertz frequencies after 100 seconds of averaging. This sensitivity allows detecting a magnetic or electric field produced by a single electron located tens of nanometers away from an NV^- center.

Using the same mechanism, the NV^- centers were employed in scanning thermal microscopy to measure high-resolution spatial maps of temperature and thermal conductivity.

Because the NV center is sensitive to magnetic fields, it is being actively used in scanning probe measurements to study myriad condensed matter phenomena both through measuring a spatially varying magnetic field or inferring local currents in a device.

Another possible use of the NV^- centers is as a detector to measure the full mechanical stress tensor in the bulk of the crystal. For this application, the stress-induced splitting of the zero-phonon-line is exploited, and its polarization properties. A robust frequency-modulated radio receiver using the electron-spin-dependent photoluminescence that operated up to 350 °C demonstrates the possibility for use in extreme conditions.

In addition to the quantum optical applications, luminescence from the NV^- centers can be applied for imaging biological processes, such as fluid flow in living cells. This application relies on good compatibility of diamond nano-particles with the living cells and on favorable properties of photoluminescence from the NV^- centers (strong intensity, easy excitation and detection, temporal stability, etc.). Compared with large single-crystal diamonds, nanodiamonds are cheap (about US$1 per gram) and available from various suppliers. Due to the relatively small size of nanodiamond, NV centers can be produced by irradiating nanodiamond of 100 nm or less with medium energy H+ beam. This method reduces the required ion dose and reaction, making it possible to mass-produce fluorescent nanodiamonds in ordinary laboratory. Fluorescent nanodiamond produced with such method is bright and photostable, making it excellent for long-term, three dimensional tracking of single particle in living cell. Those nanodiamonds are introduced in a cell, and their luminescence is monitored using a standard fluorescence microscope.

Stimulated emission from the NV^- center has been demonstrated, though it could be achieved only from the phonon side-band (i.e. broadband light) and not from the ZPL. For this purpose, the center has to be excited at a wavelength longer than ~650 nm, as higher-energy excitation ionizes the center.

The first continuous-wave room-temperature maser has been demonstrated. It used 532-nm pumped NV^- centers held within a high Purcell factor microwave cavity and an external magnetic field of 4300 G. Continuous maser oscillation generated a coherent signal at ~9.2 GHz.

The NV center can have a very long spin coherence time approaching the second regime. This is advantageous for applications in quantum sensing and quantum communication. Disadvantageous for these applications is the long radiative lifetime (~12 ns) of the NV center and the strong phonon sideband in its emission spectrum. Both issues can be addressed by putting the NV center in an optical cavity.

3.4 QUANTUM GRAVIMETERS AND GRAVITATIONAL SENSORS

Quantum gravimeters and gravitational sensors are at the forefront of modern precision measurement technology, exploiting the principles of quantum mechanics to detect gravitational forces with unprecedented sensitivity. These sensors are designed to measure local variations in gravity caused by natural or artificial changes in mass distribution, making them vital tools for applications in geophysics, civil engineering, natural resource exploration, and fundamental physics. Unlike classical gravimeters, which rely on mechanical elements like spring-mass systems, quantum gravimeters use atomic wave interference to probe gravitational fields, enhancing accuracy and reducing the influence of noise and drift. Quantum gravimeters and gravitational sensors represent a transformative leap in the science of measurement.

By leveraging the fundamental principles of quantum mechanics, these sensors provide highly accurate and stable measurements of gravitational fields, enabling a wide range of applications from resource exploration to fundamental physics. Although there are technical challenges to overcome, the progress in research and development indicates a strong trajectory toward operational maturity. As these devices continue to evolve, they will not only refine our understanding of gravity but also redefine the capabilities of sensing technologies in the 21st century and beyond.

3.4.1 The Quantum Principles behind Gravimeters

Quantum gravimeters operate based on the principles of atom interferometry, a technique in which atoms are cooled and manipulated to behave like waves. In a typical setup, a cloud of ultra-cold atoms is dropped in free fall within a vacuum chamber. As these atoms fall, they are subjected to a sequence of laser pulses that split and recombine their quantum wavefunctions. The interference pattern generated by these wavefunctions carries information about the acceleration experienced by the atoms due to gravity. Because atoms are identical and governed by the same fundamental constants, their behavior is reproducible and extremely sensitive to gravitational influences, making them ideal test particles for precise measurements.

3.4.2 From Classical to Quantum: A Paradigm Shift

Classical gravimeters, such as the LaCoste-Romberg device or superconducting gravimeters, have long served in gravity measurements. However, they suffer from several limitations, including mechanical drift, sensitivity to environmental conditions, and limited spatial resolution. The shift to quantum gravimeters marks a paradigm change in how gravitational data are acquired and interpreted. Quantum sensors eliminate many of the systemic errors found in classical devices by relying on internal atomic references rather than external mechanical ones. This not only increases stability and sensitivity but also allows for mobile and compact designs, expanding the operational environments from controlled laboratories to field-based applications.

3.4.3 Design and Operation of Quantum Gravimeters

The design of a quantum gravimeter typically involves a vacuum chamber to hold the atoms, a set of lasers for cooling and manipulating the atoms, magnetic shielding to protect the atomic states, and a detection system for reading out the interference patterns. The atoms, often rubidium or cesium, are cooled using laser cooling techniques down to microkelvin temperatures.

Once cooled, the atoms are allowed to fall freely under the influence of gravity while being subjected to a sequence of laser pulses known as a Ramsey-Bordé interferometer or a Mach-Zehnder configuration. These pulses create superpositions of momentum states, and upon recombination, the resulting interference pattern reflects the phase difference accumulated during free fall, which is directly related to gravitational acceleration.

3.4.4 Applications in Earth Observation and Geophysics

Quantum gravimeters have significant implications for Earth observation and geophysical exploration. They can detect minute changes in gravitational acceleration caused by variations in subsurface mass distributions, such as underground water reservoirs, mineral deposits, oil fields, and voids. By mapping these variations, quantum gravimeters provide valuable data for geological surveys, earthquake prediction, and environmental monitoring. For example, hydrologists can use quantum gravity data to monitor groundwater depletion or recharge, which is critical for managing water resources in arid regions. Similarly, civil engineers can employ quantum gravimetry for subsurface assessments before tunneling or construction projects to avoid potential structural hazards.

3.4.5 Role in Fundamental Physics and Metrology

Beyond practical applications, quantum gravimeters also play an important role in fundamental physics. They are used to test general relativity, search for possible violations of the equivalence principle, and probe the fine structure of spacetime. Their high precision makes them ideal for experiments involving gravitational redshift, gravitational waves, and even dark matter detection. In metrology, quantum gravimeters contribute to the redefinition of the kilogram and other SI units, ensuring that measurement standards are based on fundamental constants rather than physical artifacts. This level of precision helps maintain global consistency in scientific and industrial measurements.

3.4.6 Gravitational Gradiometry and Multi-Axis Sensing

An extension of quantum gravimetry is gravitational gradiometry, where the spatial gradient of the gravitational field is measured rather than the absolute value. This is accomplished by using two or more atom interferometers operating simultaneously at different locations or heights. The differential measurement between the sensors helps eliminate common-mode noise, such as seismic vibrations or thermal drift, and isolates the gravitational gradient. This technique is especially useful for detecting small-scale geological features and can be integrated into airborne or spaceborne platforms for regional gravity mapping. Multi-axis quantum sensors are also under development to measure gravitational vectors in multiple directions, further enhancing the capability of gravity field characterization.

3.4.7 Challenges and Technical Limitations

Despite their advantages, quantum gravimeters face several technical challenges that must be addressed for widespread adoption. The complexity of laser systems, sensitivity to magnetic fields, and the need for vacuum and cooling infrastructure pose obstacles to miniaturization and portability. Moreover, environmental noise, such as vibrations from nearby equipment or vehicles, can introduce phase errors in atom interferometry, necessitating isolation or compensation techniques. Advances in laser technology, compact vacuum systems, and integrated photonics are gradually overcoming these barriers, with recent prototypes demonstrating significant reductions in size and power consumption. Additionally, ongoing research into mobile and ruggedized systems aims to deploy quantum gravimeters in challenging terrains, such as volcanic regions, glaciers, and urban settings.

3.4.8 Integration with Other Sensor Technologies

Quantum gravimeters can be integrated with other quantum sensors, such as gyroscopes and magnetometers, to form comprehensive navigation and positioning systems. These hybrid systems have applications in autonomous navigation, where GPS signals may be unavailable, such as in underground or underwater environments. By combining gravity data with inertial measurements, these systems offer robust and drift-free navigation solutions. Furthermore, integration with satellite systems opens possibilities for global gravity monitoring, climate change studies, and earth system science. The interoperability of quantum sensors with classical data acquisition systems is also being explored to ensure seamless integration into existing infrastructure.

3.4.9 Future Prospects and Innovations

The future of quantum gravimeters is bright, with ongoing developments in miniaturization, automation, and cost reduction paving the way for commercial deployment. Governments and private companies alike are investing in quantum technology for national security, environmental monitoring, and scientific research. Emerging concepts such as chip-scale atom interferometers, space-based quantum sensors, and networked gravimeter arrays could transform our understanding of Earth's subsurface and gravitational dynamics. As artificial intelligence and machine learning techniques are applied to quantum sensor data, real-time analysis and decision-making become feasible, further enhancing the value of these sensors in critical applications.

3.5 QUANTUM ACCELEROMETERS

Quantum accelerometers represent a groundbreaking leap in precision sensing technologies, harnessing the power of quantum mechanics to measure acceleration with unprecedented accuracy. Traditional accelerometers, found in devices ranging from smartphones to aircraft, rely on mechanical or electronic systems to detect motion and orientation. While effective for many applications, classical accelerometers suffer from limitations in sensitivity, drift over time, and susceptibility to environmental disturbances. Quantum accelerometers, on the other hand, use the quantum properties of atoms—such as wave-particle duality and superposition—to provide highly stable and precise measurements that are independent of external signals like GPS.

These sensors are poised to revolutionize fields such as navigation, geophysics, and defense by offering absolute measurements of acceleration that are not subject to long-term drift. Quantum accelerometers stand at the cutting edge of sensor technology, offering a quantum leap in precision, stability, and reliability over classical counterparts. By exploiting the fundamental principles of quantum mechanics, these devices provide absolute measurements of acceleration that are essential for high-precision applications in navigation, geophysics, and defense. While technical challenges remain, the rapid pace of innovation suggests that quantum accelerometers will soon become integral components of modern sensing systems, enabling capabilities that were previously unimaginable.

3.5.1 Fundamental Principles of Operation

The core principle behind a quantum accelerometer is atom interferometry. In this technique, a cloud of ultra-cold atoms, typically rubidium or cesium, is trapped and cooled using lasers to near absolute zero. These atoms are then released into free fall in a vacuum chamber where they are manipulated using sequences of laser pulses. These laser pulses act like mirrors and beam splitters for atomic matter waves, creating a situation where the atoms travel along different paths and then interfere with each other. The resulting interference pattern encodes precise information about the acceleration the atoms experienced during their fall. Because this acceleration affects the phase of the atomic wavefunctions, it can be measured with extreme precision, leading to highly sensitive acceleration readings.

3.5.2 Comparison with Classical Accelerometers

Classical accelerometers operate using a mass-spring system or microelectromechanical systems (MEMS), where movement causes a change in capacitance, resistance, or voltage that is interpreted as acceleration. These systems, while widely used and relatively inexpensive, are subject to cumulative errors and require periodic calibration. Over time, they drift and lose accuracy, especially when used in inertial navigation systems. In contrast, quantum accelerometers provide an absolute reference based on atomic properties that are inherently stable and do not degrade with time. This means quantum devices can maintain accurate acceleration readings over long periods without external correction or recalibration. Furthermore, they are less sensitive to environmental factors like temperature changes and vibrations, which often affect classical sensors.

3.5.3 Design and Components

The design of a quantum accelerometer involves several advanced components working in unison. At the heart of the device is the vacuum chamber, where atoms are cooled and manipulated. This chamber must maintain ultra-high vacuum conditions to prevent interference from background gases. Laser systems are used to cool the atoms through Doppler and sub-Doppler cooling techniques and then to perform interferometry by manipulating the atomic states. Magnetic shielding and active stabilization systems are employed to protect the atomic wavefunctions from external noise. Finally, a detection system measures the distribution of atoms in different quantum states after interference, providing the data needed to calculate acceleration. The entire system must be isolated from mechanical vibrations to avoid noise in the measurement process.

3.5.4 Applications in Inertial Navigation

One of the most promising applications of quantum accelerometers is in inertial navigation systems, particularly in environments where GPS is unavailable or unreliable. Submarines, aircraft, and spacecraft often operate in areas where satellite signals cannot penetrate. Classical inertial navigation systems rely on integrating acceleration over time, which introduces cumulative errors. Quantum accelerometers, with their drift-free operation, offer a potential solution to this problem by providing long-term stable measurements of acceleration. When integrated with quantum gyroscopes and other quantum sensors, these accelerometers can enable fully autonomous navigation systems capable of maintaining positional accuracy over extended periods without external input.

3.5.5 Geophysical and Geological Exploration

Quantum accelerometers also have significant potential in geophysics and geological exploration. By measuring minute variations in gravitational acceleration and inertial forces, they can help detect underground structures, cavities, or mineral deposits. These capabilities are useful in oil and gas exploration, tunnel detection, and environmental monitoring. Because quantum accelerometers can be tuned to detect extremely subtle changes in movement, they are also being considered for monitoring seismic activity and tectonic shifts. Their high sensitivity could allow for earlier detection of earthquakes or more accurate mapping of geological faults, contributing to improved disaster preparedness and mitigation strategies.

3.5.6 Defense and Strategic Uses

In the defense sector, quantum accelerometers are considered game-changers for navigation, targeting, and surveillance. In GPS-denied environments, such as underwater or inside buildings, quantum inertial sensors provide a reliable alternative for determining position and movement. They are also resistant to jamming and spoofing, which are common vulnerabilities of satellite-based systems. Military vehicles, ships, and aircraft equipped with quantum accelerometers can maintain accurate navigation and targeting capabilities even when electronic warfare tactics are employed by adversaries. The strategic importance of these sensors has led many governments to invest heavily in quantum sensing research and development.

3.5.7 Challenges in Development and Deployment

Despite their enormous potential, quantum accelerometers are still in the developmental stage and face several challenges before they can be widely deployed. The current systems are bulky and require complex setups, including vacuum systems, laser arrays, and vibration isolation platforms. Miniaturizing these systems while maintaining their precision is a significant engineering challenge. Power consumption, cost, and ruggedness are also concerns that need to be addressed to make quantum accelerometers viable for field use. Advances in laser technology, microfabrication, and integrated photonics are gradually overcoming these barriers, and several prototypes have already demonstrated operation in mobile or semi-mobile environments.

3.5.8 Recent Advances and Prototypes

Research institutions and companies worldwide are working to bring quantum accelerometers from the lab to real-world applications. Notable progress includes the development of portable prototypes that can function outside of controlled laboratory environments. Some systems have already been tested in vehicles, aircraft, and marine vessels, showing promising results in navigation performance. Innovations such as chip-scale atom interferometers and fiber-based laser systems are paving the way for smaller, more robust devices. Additionally, hybrid systems combining quantum accelerometers with classical sensors are being explored to offer the best of both worlds: the reliability and maturity of classical systems with the accuracy and stability of quantum technologies.

3.5.9 Future Outlook and Impact

The future of quantum accelerometers is bright, with the potential to redefine how we perceive and measure motion. As the technology matures and becomes more accessible, it will likely be integrated into a wide range of civilian and commercial systems, from autonomous vehicles and underground navigation to earthquake monitoring and environmental mapping. Combined with artificial intelligence and data analytics, quantum sensor data could open new frontiers in science and technology. The global race to develop practical quantum sensing technologies underscores their transformative potential and strategic importance.

3.6 QUANTUM THERMOMETERS AND THERMOMETRY

Quantum thermometers represent a significant advancement in the field of temperature measurement, utilizing quantum mechanical principles to achieve levels of sensitivity and precision that classical thermometers cannot attain. In conventional thermometry, temperature is inferred from macroscopic properties such as the expansion of liquids, electrical resistance, or thermal radiation. While these methods have proven effective for most practical purposes, they face limitations in extreme conditions such as ultracold environments, nanoscale systems, or highly dynamic processes. Quantum thermometry addresses these challenges by employing quantum systems—such as trapped ions, spin states, or superconducting circuits—as temperature probes. These systems respond to temperature through changes in quantum states, enabling measurements with minimal disturbance and high spatial resolution, which are crucial for applications in quantum computing, nanotechnology, and fundamental physics. Quantum thermometry is a rapidly emerging field that leverages the unique properties of quantum systems to measure temperature with exceptional accuracy and resolution.

By transcending the limitations of classical thermometers, quantum thermometers offer new possibilities in scientific research, industrial applications, and technological innovation. Despite challenges in implementation, ongoing advancements in quantum control, sensor design, and integration are paving the way for widespread adoption of this powerful tool. As we venture deeper into the quantum realm, precise temperature measurement will remain a cornerstone of discovery and progress.

3.6.1 Principles of Quantum Thermometry

The foundation of quantum thermometry lies in the quantum behavior of particles and how thermal energy affects their states. In a quantum system, energy levels are quantized, and the population distribution across these levels is governed by the temperature of the surrounding environment.

By observing changes in this distribution, it is possible to extract precise temperature information. Quantum thermometers often use a two-level or multi-level quantum system as a probe. The population of these energy levels follows the Boltzmann distribution, which directly depends on temperature. For instance, in a two-level system, as temperature increases, the likelihood of finding the system in the excited state also increases. By preparing the system in a known quantum state and monitoring its evolution due to environmental interactions, one can infer the temperature with high accuracy.

3.6.2 Types of Quantum Thermometers

Several types of quantum thermometers have been developed, each tailored for specific measurement regimes. One common type uses atomic or ionic systems, where the population of energy states is read out through fluorescence or absorption spectroscopy. These thermometers are particularly effective at ultra-low temperatures where classical methods fail. Another variant involves superconducting qubits, which are sensitive to temperature-induced decoherence. As temperature rises, the coherence of the quantum state diminishes, providing a measurable indicator.

Quantum dots and nitrogen-vacancy (NV) centers in diamond also serve as versatile quantum thermometers. NV centers, for instance, are defects in the diamond lattice that can be manipulated with optical and microwave fields. They are extremely sensitive to their environment, including temperature, magnetic fields, and strain, making them suitable for high-resolution thermal imaging at the nanoscale.

3.6.3 Advantages over Classical Thermometry

Quantum thermometers offer numerous advantages over classical approaches, especially in terms of sensitivity, spatial resolution, and non-invasiveness. Classical thermometers often require direct contact with the object being measured, which can be problematic in small or delicate systems. In contrast, quantum thermometers can operate remotely or with minimal interaction, reducing the risk of perturbing the system. Additionally, classical sensors struggle to measure temperature variations at the nanometer scale or at very low temperatures where thermal energy is extremely limited. Quantum thermometers, by leveraging the discrete nature of quantum states and their interaction with thermal environments, can provide accurate readings even in such extreme conditions. This level of sensitivity is crucial for studying quantum devices, biological cells, or materials under cryogenic conditions.

3.6.4 Applications in Quantum Technologies

Quantum thermometry plays a critical role in the development and operation of quantum technologies. In quantum computing, for example, maintaining coherence requires precise thermal control. Even slight temperature fluctuations can cause decoherence, leading to errors in computation. Quantum thermometers are used to monitor and stabilize the temperature of qubit systems, ensuring optimal performance. In cryogenics, quantum thermometry provides a means to verify cooling processes down to millikelvin or even nanokelvin temperatures, where conventional sensors lose reliability. These measurements are vital for experiments in quantum simulation, condensed matter physics, and fundamental tests of quantum mechanics. Moreover, in nanoscale heat transport studies, quantum thermometers enable the observation of thermal gradients and heat flow with unprecedented resolution, offering insights into thermodynamic behavior at the microscopic level.

3.6.5 Techniques and Readout Methods

The implementation of quantum thermometry requires sophisticated techniques to prepare, control, and read out quantum states. Optical techniques are widely used, particularly for systems like NV centers or cold atoms. For instance, changes in fluorescence intensity or spectral shifts provide temperature-dependent signals. In other cases, quantum coherence and entanglement are exploited. Ramsey interferometry and Rabi oscillations are techniques where the evolution of a quantum state over time reveals temperature information through phase accumulation or frequency changes. These methods require precise control over the quantum system and often involve complex laser or microwave systems. The readout process must also be fast and accurate, as thermal noise and decoherence can quickly degrade the quantum state. Advances in photonics and quantum control have made these processes increasingly feasible in laboratory and applied settings.

3.6.6 Challenges and Limitations

Despite their promise, quantum thermometers face challenges in practicality and scalability. One major issue is the complexity of the experimental setup, which often requires vacuum systems, cryogenic cooling, and precise alignment of optical components. These requirements limit the deployment of quantum thermometry outside laboratory environments. Additionally, quantum sensors are sensitive not only to temperature but also to other environmental variables such as magnetic and electric fields, which can complicate data interpretation. Efforts are ongoing to improve the robustness and portability of quantum thermometers, including the development of integrated sensor platforms and miniaturized quantum devices. Another challenge is calibration, as quantum thermometers rely on theoretical models to relate quantum state changes to temperature. Accurate modeling and benchmarking against known standards are essential to ensure measurement fidelity.

3.6.7 Emerging Directions and Innovations

Recent developments in quantum thermometry are focusing on enhancing spatial resolution, increasing robustness, and integrating with other quantum sensing technologies. One promising direction is quantum-enhanced thermometry using entangled states, which can surpass classical limits of precision known as the standard quantum limit. By entangling multiple quantum sensors, it is possible to reduce uncertainty and achieve Heisenberg-limited sensitivity. Another area of interest is the use of hybrid systems, such as coupling NV centers to photonic or plasmonic structures to enhance signal strength and thermal responsiveness. Researchers are also exploring biological applications, such as intracellular thermometry, where quantum sensors are introduced into living cells to monitor temperature dynamics in real time. These applications have the potential to revolutionize biomedical diagnostics and therapy by providing thermal maps at the cellular or even molecular level.

3.6.8 Future Prospects and Impact

The future of quantum thermometry lies in its integration into broader scientific and technological frameworks. As quantum computing and nanotechnology continue to advance, the demand for precise thermal measurement will grow. Quantum thermometers will be essential tools for ensuring the stability and reliability of these systems. Furthermore, their application in materials science, chemistry, and biology opens up new avenues for exploration and innovation.

From mapping heat dissipation in microprocessors to tracking metabolic processes in single cells, the ability to measure temperature with quantum precision is poised to have a transformative impact. Continued research and investment in quantum technologies will accelerate the development of practical, deployable quantum thermometers that can be used across various disciplines.

3.7 QUANTUM PHOTONIC SENSORS

Quantum photonic sensors represent a remarkable intersection of quantum mechanics and optical technology. These sensors exploit the quantum properties of photons—such as superposition, entanglement, and squeezing—to perform measurements with extreme sensitivity, resolution, and precision. Unlike classical optical sensors that rely on the intensity or phase of light alone, quantum photonic sensors utilize the quantum state of light to detect subtle changes in the environment. This capability makes them highly suitable for applications in metrology, navigation, biological imaging, environmental monitoring, and fundamental physics. As research in quantum technologies advances, quantum photonic sensors are rapidly becoming vital tools for pushing the frontiers of precision measurement. Quantum photonic sensors represent a transformative advancement in the science of measurement. By harnessing the unique properties of quantum light, these sensors achieve levels of precision and sensitivity that surpass the capabilities of classical optical systems. Their applications span across diverse fields—from health care and environmental monitoring to aerospace and fundamental physics. While technical challenges remain, ongoing research and technological innovation are steadily overcoming these hurdles, paving the way for broader adoption. As quantum photonic technologies continue to mature and integrate with existing systems, their impact on science, industry, and society is expected to be profound and far-reaching.

3.7.1 Fundamental Concepts of Quantum Photonics

At the heart of quantum photonic sensing lies the manipulation of individual photons or entangled photon pairs. Photons are ideal carriers of quantum information because they interact weakly with their environment, which helps preserve coherence over long distances and time scales. In a quantum photonic sensor, information about a target quantity—such as displacement, magnetic field, or refractive index—is encoded in specific quantum features of light. Quantum states like squeezed states reduce uncertainty in one observable (like phase) at the cost of increasing it in the conjugate variable (like amplitude), allowing the sensor to exceed the classical shot-noise limit. Entangled photons, on the other hand, share correlations that remain intact even when separated by large distances, enabling measurements with higher sensitivity and security.

3.7.2 Design and Operation of Quantum Photonic Sensors

Quantum photonic sensors are designed to maximize the interaction between light and the quantity being measured while preserving the delicate quantum properties of photons. Typically, a source generates quantum light, such as entangled photon pairs through spontaneous parametric down-conversion or squeezed states using nonlinear optical media. These quantum states are then sent through a sensing interferometer or waveguide system.

The quantity of interest—whether it be a physical displacement, pressure variation, or temperature change—causes a phase shift or alteration in the light path, which is then read out by single-photon detectors or homodyne measurement systems. Since quantum states are highly sensitive to phase changes, even minute environmental alterations can be detected with great precision.

3.7.3 Advantages over Classical Optical Sensors

Quantum photonic sensors surpass classical optical sensors in several key areas, primarily sensitivity and noise resilience. While classical sensors are fundamentally limited by the standard quantum limit, quantum sensors can reach or approach the Heisenberg limit, which defines the ultimate bound of measurement precision dictated by quantum mechanics.

This enables a higher signal-to-noise ratio, especially in low-light conditions or when detecting weak signals. Moreover, quantum photonic sensors are less affected by certain types of classical noise, such as thermal or electronic fluctuations, due to the non-classical nature of their signal processing. These advantages are particularly beneficial in high-precision applications like gravitational wave detection, atomic force microscopy, and quantum-enhanced medical diagnostics.

3.7.4 Applications in Biomedical and Environmental Fields

One of the most promising application areas for quantum photonic sensors is biomedical imaging and diagnostics. Techniques such as quantum optical coherence tomography and quantum microscopy employ entangled photons to achieve higher resolution and contrast while reducing the radiation dose delivered to biological tissues.

These techniques are invaluable for imaging delicate samples like neural tissues or embryonic cells. In environmental science, quantum photonic sensors are used for the detection of trace gases, pollutants, and temperature changes with a precision unmatched by classical devices. Their ability to detect single-photon level signals makes them ideal for monitoring greenhouse gases or pollutants in remote and low-light environments, contributing to better climate modeling and ecological protection.

3.7.5 Quantum Interferometry in Sensing

Quantum interferometry is a cornerstone technique in quantum photonic sensing. Interferometers, such as the Mach-Zehnder or Michelson types, are used to measure phase shifts induced by external influences. When quantum states like NOON states—where photons are path-entangled in superposition—are used in these interferometers, the phase sensitivity can exceed classical limits. This allows for ultra-precise measurements of distances, accelerations, and rotations. In fundamental physics, such interferometry has been used to test the limits of general relativity and detect gravitational waves, as in the case of the LIGO experiment where squeezed light was incorporated to enhance sensitivity. Similarly, quantum-enhanced gyroscopes and accelerometers based on photonic principles are being developed for use in navigation systems where GPS is unavailable.

3.7.6 Integration with Photonic Circuits and Technologies

The development of integrated photonic circuits has significantly enhanced the practicality of quantum photonic sensors. On-chip implementations reduce the size, weight, and power consumption of these systems, making them viable for field applications and commercial deployment. Integrated circuits also improve stability and allow for the fabrication of complex sensor networks that can operate in parallel. Silicon photonics, lithium niobate platforms, and diamond photonics are some of the materials being explored for integrating quantum photonic components, including sources, interferometers, modulators, and detectors. These advances open the door to scalable quantum sensor arrays that can be deployed in aerospace, defense, telecommunications, and even consumer electronics.

3.7.7 Challenges in Quantum Photonic Sensing

Despite their advantages, quantum photonic sensors face several technical and practical challenges. Generating and maintaining high-quality quantum states of light requires sophisticated equipment and controlled environments. Photon loss, decoherence, and detector inefficiencies can degrade sensor performance, especially in noisy or uncontrolled settings. Scaling up the number of entangled photons or squeezed modes while preserving quantum coherence is another complex task that limits the resolution and reliability of current systems. Moreover, the readout mechanisms for quantum sensors often require single-photon detectors that are expensive and cryogenically cooled, though newer room-temperature alternatives are under development. Addressing these challenges is essential for transitioning quantum photonic sensors from laboratory research to widespread application.

3.7.8 Future Directions and Innovations

The field of quantum photonic sensing is evolving rapidly, driven by breakthroughs in quantum optics, materials science, and nanofabrication. Emerging trends include the use of machine learning for quantum state estimation and sensor optimization, which can further enhance precision and robustness. Hybrid systems that combine quantum photonic sensors with other quantum technologies, such as atomic sensors or superconducting devices, are also being explored to achieve multi-parameter sensing. The advent of space-based quantum sensing missions promises to extend the reach of these technologies to planetary exploration, space weather monitoring, and satellite-based navigation systems. Governments and industry leaders are investing heavily in quantum sensing research, recognizing its potential to revolutionize a wide range of scientific and industrial domains.

CHAPTER

Quantum Sensing Techniques

4

4.1 QUANTUM INTERFEROMETRY FOR PRECISION MEASUREMENT

Quantum interferometry is a powerful and refined technique that leverages quantum mechanical principles to achieve extraordinarily precise measurements. It represents the integration of interferometry—traditionally an optical method for measuring small displacements, refractive index changes, and surface irregularities—with the unique properties of quantum states of matter and light. Classical interferometers, such as the Michelson or Mach-Zehnder types, utilize the interference of light waves to measure phase shifts caused by physical phenomena. Quantum interferometry takes this a step further by employing quantum states such as entangled photons, squeezed light, or matter waves to surpass the classical limits of measurement precision, thus offering significant advantages in fields ranging from gravitational wave detection to medical imaging. Quantum interferometry represents a transformative advance in the science of precision measurement. By harnessing the subtle and powerful effects of quantum superposition, entanglement, and interference, it enables levels of sensitivity far beyond classical capabilities. From exploring the universe through gravitational wave detection to revolutionizing biomedical imaging and environmental monitoring, the reach of quantum interferometry continues to grow. While technical challenges remain, the trajectory of research and technological development strongly indicates that quantum interferometry will play a foundational role in the next era of scientific discovery and technological innovation.

4.1.1 Principles of Quantum Interference

The core principle behind quantum interferometry is the phenomenon of quantum interference. This occurs when quantum particles like photons or atoms exist in a superposition of states and interfere with themselves upon recombination. In classical optics, interference patterns result from the coherent addition of electromagnetic waves. In contrast, quantum interference arises due to the probability amplitudes of different quantum paths combining constructively or destructively. This allows for phase sensitivity enhancements not achievable with classical light. By carefully preparing quantum states and manipulating their evolution in an interferometric setup, researchers can measure phase shifts with sensitivities approaching or reaching the Heisenberg limit, which is the ultimate bound allowed by quantum mechanics.

4.1.2 Quantum States in Interferometry

To exploit the full potential of quantum interferometry, special quantum states of light or matter are used. One such example is the NOON state, where N particles are simultaneously in one path and zero in the other, existing in a superposition of all particles being in either path. These states provide phase sensitivity that scales with $1/N$, surpassing the classical shot-noise limit of $1/\sqrt{N}$. Another important quantum state is the squeezed state, which reduces uncertainty in one variable, such as phase, at the expense of increased uncertainty in the conjugate variable, like amplitude. Squeezed light, in particular, is commonly used in interferometers to reduce quantum noise and enhance signal detection, particularly in low-light or high-noise environments.

4.1.3 Applications in Gravitational Wave Detection

One of the most striking real-world applications of quantum interferometry is in gravitational wave observatories such as LIGO and Virgo. These facilities use highly sensitive Michelson interferometers to detect minute distortions in spacetime caused by distant cosmic events like black hole mergers. However, even with kilometer-long arms and vibration isolation, classical noise imposes a fundamental limit on sensitivity. By injecting squeezed vacuum states into the interferometer's unused port, researchers have been able to reduce quantum shot noise and enhance sensitivity. This integration of quantum techniques has significantly improved the detection rate and accuracy of gravitational wave measurements, marking a major milestone in both astrophysics and quantum technology.

4.1.4 Quantum Interferometry with Atoms and Ions

In addition to photonic systems, quantum interferometry can be implemented with massive particles such as atoms or ions. Matter-wave interferometry uses the wave-like nature of atoms, often cooled to near absolute zero temperatures and trapped using magnetic or optical fields. These atomic interferometers are extremely sensitive to inertial forces and are being explored for applications in precision navigation, seismology, and tests of general relativity. Atom interferometers split and recombine matter waves, with any acceleration or gravitational change along their paths causing a measurable phase shift. Such systems can detect changes in gravitational fields with extraordinary precision, making them ideal for geophysical surveys and underground structure detection.

4.1.5 Quantum Metrology and the Heisenberg Limit

Quantum interferometry is a cornerstone of the broader field of quantum metrology, which aims to use quantum phenomena to improve measurement precision. The standard quantum limit, dictated by the statistical nature of quantum measurements, sets a bound on the precision achievable with classical resources. However, by using entangled states or squeezed states, quantum metrology enables measurements at or near the Heisenberg limit, where precision scales inversely with the number of particles used. This dramatic improvement has profound implications for timekeeping, frequency standards, and fundamental constant measurements. Atomic clocks using quantum interferometry techniques now define international time standards with uncertainties of less than one second over billions of years.

4.1.6 Challenges and Technical Barriers

Despite its promise, quantum interferometry faces significant technical challenges. Preparing and maintaining fragile quantum states over the duration of a measurement requires stringent isolation from environmental noise. Photon loss, decoherence, and thermal fluctuations can degrade the quality of interference and reduce sensitivity. Moreover, generating entangled or squeezed states at scale remains a complex and resource-intensive task. In atom interferometry, maintaining coherence over macroscopic distances and dealing with systematic effects like magnetic field fluctuations or laser phase noise also present major hurdles. Advances in quantum control, error correction, and robust device engineering are essential to overcoming these limitations and realizing the full potential of quantum interferometry in practical settings.

4.1.7 Quantum Interferometry in Imaging and Sensing

Beyond fundamental physics, quantum interferometry is making inroads into imaging and sensing applications. Quantum-enhanced optical coherence tomography (OCT) uses entangled photon pairs to achieve higher image resolution with lower photon doses, which is critical in sensitive biological systems such as retinal imaging. Similarly, interferometric setups using quantum light can detect refractive index changes, displacement, or temperature variations at the nanoscale, making them valuable tools in materials science and nanotechnology. In magnetometry, quantum interferometers detect minute magnetic field changes with sensitivity orders of magnitude better than classical sensors, enabling applications in medical diagnostics such as magnetoencephalography and magnetocardiography.

4.1.8 Integration and Miniaturization of Quantum Interferometers

Recent advancements in integrated photonics and nanofabrication have made it possible to build compact, chip-scale quantum interferometers. These devices integrate light sources, interferometric circuits, and detectors on a single platform, enabling scalable and robust quantum sensors. Silicon photonics and other materials like lithium niobate or diamond are being used to fabricate these integrated systems. Miniaturized quantum interferometers are not only more portable and less sensitive to environmental disturbances but also offer the potential for mass production and deployment in diverse environments. This opens up possibilities for deploying quantum sensors in autonomous vehicles, aerospace systems, and mobile health devices.

4.1.9 Future Outlook of Quantum Interferometry

The future of quantum interferometry lies in continued innovation across multiple fronts: quantum state engineering, integrated quantum optics, improved materials, and enhanced computational models. As research progresses, we can expect quantum interferometers to become central components in a new generation of ultra-precise measurement tools. Potential developments include entanglement-enhanced radar, space-based gravitational mapping, and quantum networks for distributed sensing. Combining quantum interferometry with machine learning and artificial intelligence may also lead to self-calibrating, adaptive measurement systems that automatically optimize sensitivity in real-time. Furthermore, as governments and industries invest heavily in quantum technologies, the commercial applications of quantum interferometry are likely to expand, influencing everything from navigation and infrastructure to communications and defense.

4.2 SENSING USING ENTANGLED STATES

Quantum entanglement is one of the most fascinating and non-intuitive phenomena in quantum mechanics, offering capabilities far beyond those of classical physics. In the realm of quantum sensing, entangled states serve as a fundamental resource for achieving sensitivities and accuracies that classical systems cannot match. Entanglement refers to a quantum correlation between particles such that the state of one particle cannot be described independently of the state of the other, regardless of the distance between them. When employed in sensing applications, these correlations allow for enhanced measurement precision by leveraging collective quantum behavior. Entangled states can be utilized to measure a variety of physical quantities, such as time, magnetic fields, acceleration, or gravitational forces, with an accuracy that surpasses traditional limits.

Sensing using entangled states stands at the forefront of quantum technology, offering transformative potential in terms of measurement precision, sensitivity, and novel functionality. By utilizing the deep quantum correlations present in entangled systems, it becomes possible to surpass classical measurement limits and open new frontiers in metrology, imaging, navigation, and fundamental science. While challenges in scalability, coherence, and state preparation remain, rapid progress in experimental techniques and theoretical understanding continues to bring this vision closer to reality. As quantum technologies mature, entangled-state-based sensing is set to revolutionize how we observe and interact with the physical world, reshaping fields as diverse as medicine, astronomy, and national security.

4.2.1 Theoretical Basis of Entanglement-Enhanced Sensing

At the core of entangled state-based sensing is the idea that entanglement reduces the uncertainty associated with certain measurements. In conventional or classical sensing systems, the precision of a measurement is generally limited by the standard quantum limit, which scales with the inverse square root of the number of particles or photons used. This limitation arises due to shot noise, a form of quantum uncertainty inherent to uncorrelated particles. However, entangled states allow the system to approach or even reach the Heisenberg limit, where the measurement uncertainty scales inversely with the number of particles. This means that a sensor based on entangled particles can perform significantly better than one based on independent particles. Theoretically, this opens the door to ultra-sensitive instruments capable of detecting minute changes in physical parameters with unprecedented accuracy.

4.2.2 Types of Entangled States Used in Sensing

Different types of entangled states are employed depending on the application and the nature of the measurement. One of the most commonly used forms is the Greenberger-Horne-Zeilinger (GHZ) state, which is a maximally entangled state involving multiple particles. GHZ states are particularly useful in interferometric setups and have been shown to enhance phase estimation. Another important class of entangled states used in sensing is the NOON state, in which all particles are in a superposition of being in one mode or another. NOON states are highly sensitive to phase shifts and are extensively used in optical and atomic interferometry. Additionally, spin-squeezed states are partially entangled states where the quantum noise is redistributed in a way that reduces uncertainty in the measured quantity. These states are particularly effective in applications such as atomic clocks and magnetometers.

4.2.3 Quantum Entanglement in Atomic Clocks

One of the most successful implementations of entangled states in sensing can be found in atomic clocks. Traditional atomic clocks rely on transitions between energy levels of atoms to measure time with high precision. However, their performance is constrained by the quantum projection noise of uncorrelated atoms. By entangling the atoms in the clock, scientists have demonstrated a significant reduction in noise and a corresponding improvement in timekeeping precision. These advancements not only improve the accuracy of global navigation satellite systems but also facilitate more precise measurements of fundamental physical constants. Entangled atomic ensembles in clocks represent a practical demonstration of how quantum resources can be harnessed for real-world sensing improvements.

4.2.4 Entangled Photons in Quantum Imaging

In the field of imaging, entangled photons have been used to develop techniques that allow for higher resolution and reduced noise. Quantum imaging methods such as ghost imaging and quantum lithography exploit the spatial correlations of entangled photon pairs to reconstruct images even when one of the photons does not interact directly with the object being imaged. These techniques offer potential applications in biological imaging, where low light levels are essential to prevent sample damage. Moreover, entangled photons can improve phase sensitivity in interferometric measurements, enabling imaging of structures at the nanoscale that would otherwise be beyond the reach of classical optical systems.

4.2.5 Quantum Magnetometry Using Entanglement

Quantum magnetometers that employ entangled states have been shown to measure magnetic fields with unprecedented sensitivity. These sensors are based on atomic ensembles where the spins of the atoms are entangled in a way that enhances their collective response to magnetic fields. By minimizing the quantum noise in the spin direction sensitive to the magnetic field, these magnetometers can detect extremely weak magnetic signals.

This capability is particularly valuable in biomedical applications, such as detecting neural activity in the brain through magnetoencephalography or monitoring cardiac signals. Furthermore, quantum magnetometers have potential use in geological surveys and detecting unexploded ordnance by identifying magnetic anomalies in the environment.

4.2.6 Challenges in Generating and Maintaining Entangled States

Despite the promising advantages, the practical implementation of entangled states in sensing applications faces significant challenges. Generating entangled states, especially for a large number of particles, is technically demanding and often requires precise control over experimental conditions. Additionally, entangled states are highly susceptible to decoherence, a process by which the system loses its quantum correlations due to interaction with the environment. This makes it difficult to maintain entanglement over extended periods or in noisy, real-world conditions. Another challenge lies in the efficient detection of entangled states, as current measurement techniques can be invasive or inefficient, thus reducing the overall sensitivity gains. Overcoming these hurdles requires advances in quantum error correction, improved isolation techniques, and better state preparation methods.

4.2.7 Entanglement in Quantum Networks and Distributed Sensing

Entangled states also open possibilities for distributed quantum sensing, where multiple sensors are entangled across large distances to function as a coherent sensing network. Such networks could be used to detect spatially varying fields with high resolution or to perform correlated measurements that would be impossible using independent sensors.

For instance, entangled sensors placed at different locations on the Earth or in space could monitor gravitational gradients or seismic activity with enhanced precision. Quantum networks employing entanglement may also contribute to secure communication and synchronization in sensitive military or scientific operations, making entanglement a valuable asset in a variety of strategic domains.

4.2.8 Emerging Technologies and Future Outlook

Recent advancements in photonics, quantum optics, and nanotechnology are paving the way for more practical and scalable quantum sensors based on entangled states. Integrated quantum photonic circuits can now generate, manipulate, and measure entangled photons on a single chip, reducing system complexity and increasing robustness. In parallel, hybrid systems that combine solid-state qubits such as nitrogen-vacancy centers in diamond with optical entanglement are showing promise for field-deployable sensors. Looking forward, the development of quantum repeaters and long-lived quantum memories will enhance the viability of entangled-state-based sensing over large distances. With growing interest from both academia and industry, quantum sensing using entangled states is likely to play a transformative role in precision measurement technologies of the future.

4.3 QUANTUM STATE TOMOGRAPHY IN SENSING

Quantum tomography or quantum state tomography is the process by which a quantum state is reconstructed using measurements on an ensemble of identical quantum states. The source of these states may be any device or system which prepares quantum states either consistently into quantum pure states or otherwise into general mixed states. To be able to uniquely identify the state, the measurements must be tomographically complete. That is, the measured operators must form an operator basis on the Hilbert space of the system, providing all the information about the state. Such a set of observations is sometimes called a quorum. The term *tomography* was first used in the quantum physics literature in a 1993 paper introducing experimental optical homodyne tomography.

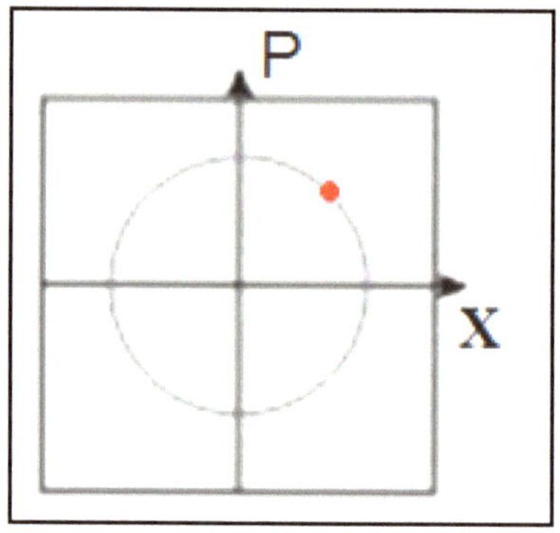

Fig: One Harmonic Oscillator Represented in the Phase Space by its Momentum and Position

In quantum process tomography on the other hand, known quantum states are used to probe a quantum process to find out how the process can be described. Similarly, quantum measurement tomography works to find out what measurement is being performed. Whereas, randomized benchmarking scalably obtains a figure of merit of the overlap between the error prone physical quantum process and its ideal counterpart.

The general principle behind quantum state tomography is that by repeatedly performing many different measurements on quantum systems described by identical density matrices, frequency counts can be used to infer probabilities, and these probabilities are combined with Born's rule to determine a density matrix which fits the best with the observations.

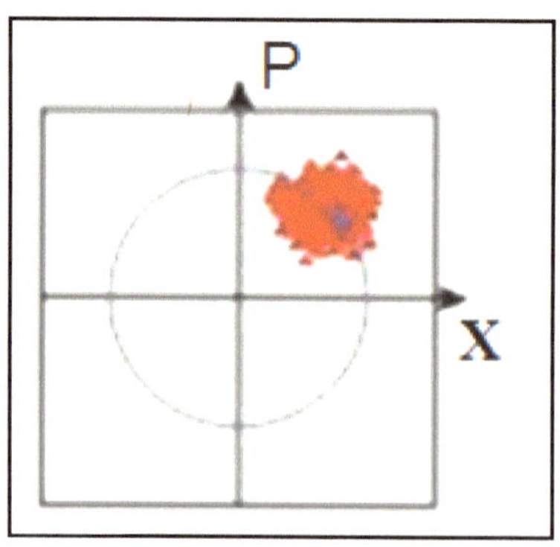

Fig: Many Identical Oscillators Represented in the Phase Space by their Momentum and Position

This can be easily understood by making a classical analogy. Consider a harmonic oscillator (e.g. a pendulum). The position and momentum of the oscillator at any given point can be measured and therefore the motion can be completely described by the phase space. This is shown in figure. By performing this measurement for a large number of identical oscillators we get a probability distribution in the phase space. This distribution can be normalized (the oscillator at a given time has to be somewhere) and the distribution must be non-negative. So we have retrieved a function $W(x, p)$ which gives a description of the chance of finding the particle at a given point with a given momentum.

For quantum mechanical particles the same can be done. The only difference is that the Heisenberg's uncertainty principle mustn't be violated, meaning that we cannot measure the particle's momentum and position at the same time. The particle's momentum and its position are called quadratures in quantum related states. By measuring one of the quadratures of a large number of identical quantum states will give us a probability density corresponding to that particular quadrature. This is called the marginal distribution, $\mathrm{pr}(X)\, or\, \mathrm{pr}(P)$.

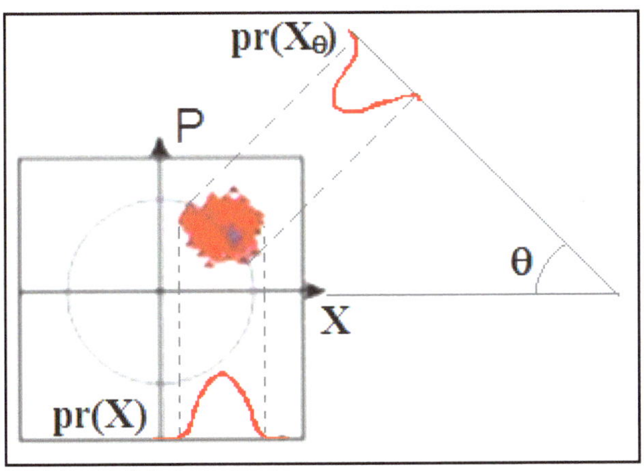

Fig: Marginal Distribution

4.3.1 What Quantum State Tomography is Used

Quantum tomography is applied on a source of systems, to determine the quantum state of the output of that source. Unlike a measurement on a single system, which determines the system's current state after the measurement (in general, the act of making a measurement alters the quantum state), quantum tomography works to determine the state(s) prior to the measurements.

Quantum tomography can be used for characterizing optical signals, including measuring the signal gain and loss of optical devices, as well as in quantum computing and quantum information theory to reliably determine the actual states of the qubits. One can imagine a situation in which a person Bob prepares many identical objects (particles or fields) in the same quantum states and then gives them to Alice to measure. Not confident with Bob's description of the state, Alice may wish to do quantum tomography to classify the state herself.

4.3.2 Methods of Quantum State Tomography

Linear Inversion

Using Born's rule, one can derive the simplest form of quantum tomography. Generally, being in a pure state is not known in advance, and a state may be mixed. In this case, many different types of measurements will have to be performed, many times each. To fully reconstruct the density matrix for a mixed state in a finite-dimensional Hilbert space, the following technique may be used.

Born's rule states $\mathrm{P}(E_i \mid \rho) = \mathrm{Trace}(E_i \rho)$, where E_i is a particular measurement outcome projector and ρ is the density matrix of the system. Given a histogram of observations for each measurement, one has an approximation p_i to $\mathrm{P}(E_i \mid \rho)$ for each E_i.

Given linear operators S S and T T, define the inner product

$$S \cdot T = \mathrm{Tr}[S^{\dagger} T] = \vec{S}^{\dagger} \vec{T}$$

Define the matrix A A as

$$A = \begin{pmatrix} \vec{E}_1^{\dagger} \\ \vec{E}_2^{\dagger} \\ \vec{E}_3^{\dagger} \\ \vdots \end{pmatrix}$$

Here E_i is some fixed list of individual measurements (with binary outcomes), and A does all the measurements at once. Then applying this to $\vec{\rho}$ yields the probabilities:

$$A\vec{\rho} = \begin{pmatrix} \vec{E}_1^{\dagger}\vec{\rho} \\ \vec{E}_2^{\dagger}\vec{\rho} \\ \vec{E}_3^{\dagger}\vec{\rho} \\ \vdots \end{pmatrix} = \begin{pmatrix} E_1 \cdot \rho \\ E_2 \cdot \rho \\ E_3 \cdot \rho \\ \vdots \end{pmatrix} = \begin{pmatrix} \mathrm{P}(E_1 \mid \rho) \\ \mathrm{P}(E_2 \mid \rho) \\ \mathrm{P}(E_3 \mid \rho) \\ \vdots \end{pmatrix} \approx \begin{pmatrix} p_1 \\ p_2 \\ p_3 \\ \vdots \end{pmatrix} = \vec{p}$$

This system is not going to be square in general, as for each measurement being made there will generally be multiple measurement outcome projectors E_i. For example, in a 2-D Hilbert space with 3 measurements $\sigma_x, \sigma_y, \sigma_z$, each measurement has 2 outcomes, each of which has a projector E_i, for 6 projectors, whereas the real dimension of the space of density matrices is $(2 \cdot 2^2)/2=4$, leaving A A to be 6 x 4. To solve the system, multiply on the left by A^T:

$$A^T A \vec{\rho} = A^T \vec{p}$$

Now solving for $\vec{\rho}$ yields the pseudoinverse:

$$\vec{\rho} = (A^T A)^{-1} A^T \vec{p}$$

This works in general only if the measurement list E_i is tomographically complete. Otherwise, the matrix $A^T A$ will not be invertible.

Continuous Variables and Quantum Homodyne Tomography

In infinite dimensional Hilbert spaces, e.g. in measurements of continuous variables such as position, the methodology is somewhat more complex. One notable example is in the tomography of light, known as optical homodyne tomography. Using balanced homodyne measurements, one can derive the Wigner function and a density matrix for the state of the light. One approach involves measurements along different rotated directions in phase space.

For each direction θ, one can find a probability distribution $w(q,\theta)$ for the probability density of measurements in the θ direction of phase space yielding the value q q. Using an inverse Radon transformation (the filtered back projection) on $w(q,\theta)$ leads to the Wigner function, $W(x,p)$, which can be converted by an inverse Fourier transform into the density matrix for the state in any basis. A similar technique is often used in medical tomography.

Example: Single-Qubit State Tomography

The density matrix of a single qubit can be expressed in terms of its Bloch vecto $\vec{r}$ and the Pauli vector $\vec{\sigma}$:

$$\rho = \frac{1}{2}(I + \vec{r} \cdot \vec{\sigma}) = \frac{1}{2}\begin{pmatrix} 1+r_z & r_x - \mathrm{i}r_y \\ r_x + \mathrm{i}r_y & 1-r_z \end{pmatrix}$$

The single-qubit state tomography can be performed by means of single-qubit Pauli measurements:

- First, create a list of three quantum circuits, with the first one measuring the qubit in the computational basis (Z-basis), the second one performing a Hadamard gate before measurement (which makes the measurement in X-basis), and the third one performing the appropriate phase shift gate (that is $\sqrt{Z}^{\dagger} = |0\rangle\langle 0| + \exp(-\mathrm{i}\pi/2)|1\rangle\langle 1|$) followed by a Hadamard gate before measurement (which makes the measurement in Y-basis);
- Then, run these circuits (typically thousands of times), and the counts in the measurement results of the first circuit produces $\overline{z} = (n_{z,+} - n_{z,-})/(n_{z,+} + n_{z,-})$, the second circuit $\overline{x}$, and the third circuit $\overline{y}$;
- Finally, if $\overline{x}^2 + \overline{y}^2 + \overline{z}^2 \le 1$, then a measured Bloch vector is produced as $\vec{r}_m = (\overline{x}, \overline{y}, \overline{z})$, and the measured density matrix is $\rho_m = \frac{1}{2}(I + \vec{r}_m \cdot \vec{\sigma})$; If $\overline{x}^2 + \overline{y}^2 + \overline{z}^2 > 1$ it'll be necessary to renormalize the measured Bloch

vector as $\vec{r}_m = (\bar{x}, \bar{y}, \bar{z}) / \sqrt{\bar{x}^2 + \bar{y}^2 + \bar{z}^2}$ before using it to calculate the measured density matrix.

This algorithm is the foundation for qubit tomography and is used in some quantum programming routines, like that of Qiskit.

Example: Homodyne Tomography

Electromagnetic field amplitudes (quadratures) can be measured with high efficiency using photodetectors together with temporal mode selectivity. Balanced homodyne tomography is a reliable technique of reconstructing quantum states in the optical domain. This technique combines the advantages of the high efficiencies of photodiodes in measuring the intensity or photon number of light, together with measuring the quantum features of light by a clever set-up called the homodyne tomography detector.

Quantum homodyne tomography is understood by the following example. A laser is directed onto a 50-50% beamsplitter, splitting the laser beam into two beams. One is used as a local oscillator (LO) and the other is used to generate photons with a particular quantum state. The generation of quantum states can be realized, e.g. by directing the laser beam through a frequency doubling crystal and then onto a parametric down-conversion crystal.

This crystal generates two photons in a certain quantum state. One of the photons is used as a trigger signal used to trigger (start) the readout event of the homodyne tomography detector. The other photon is directed into the homodyne tomography detector, in order to reconstruct its quantum state. Since the trigger and signal photons are entangled, it is important to realize that the optical mode of the signal state is created nonlocal only when the trigger photon impinges the photodetector (of the trigger event readout module) and is actually measured. More simply said, it is only when the trigger photon is measured, that the signal photon can be measured by the homodyne detector.

Now consider the homodyne tomography detector as depicted in figure 4 (figure missing). The signal photon (this is the quantum state we want to reconstruct) interferes with the local oscillator, when they are directed onto a 50-50% beamsplitter. Since the two beams originate from the same so called master laser, they have the same fixed phase relation. The local oscillator must be intense, compared to the signal so it provides a precise phase reference. The local oscillator is so intense, that we can treat it classically ($a = \alpha$) and neglect the quantum fluctuations.

The signal field is spatially and temporally controlled by the local oscillator, which has a controlled shape. Where the local oscillator is zero, the signal is rejected. Therefore, we have temporal-spatial mode selectivity of the signal. The beamsplitter redirects the two beams to two photodetectors. The photodetectors generate an electric current proportional to the photon number. The two detector currents are subtracted and the resulting current is proportional to the electric field operator in the signal mode, depended on relative optical phase of signal and local oscillator.

Since the electric field amplitude of the local oscillator is much higher than that of the signal the intensity or fluctuations in the signal field can be seen. The homodyne tomography system functions as an amplifier. The system can be seen as an interferometer with such a high intensity reference beam (the local oscillator) that unbalancing the interference by a single photon in the signal is measurable.

The measurement is reproduced a large number of times. Then the phase difference between the signal and local oscillator is changed in order to 'scan' a different angle in the phase space. This can be seen from figure 4. The measurement is repeated again a large number of times and a marginal distribution is retrieved from the current difference. The marginal distribution can be transformed into the density matrix and/or the Wigner function. Since the density matrix and the Wigner function give information about the quantum state of the photon, we have reconstructed the quantum state of the photon. The advantage of this balanced detection method is that this arrangement is insensitive to fluctuations in the intensity of the laser. The quantum computations for retrieving the quadrature component from the current difference are performed as follows.

The photon number operator for the beams striking the photodetectors after the beamsplitter is given by:

$$\hat{n}_i = \hat{a}_i^\dagger \hat{a}_i$$

where i is 1 and 2, for respectively beam one and two. The mode operators of the field emerging the beamsplitters are given by:

$$\hat{a}_1 = 2^{-1/2}(\hat{a} - \alpha_{LO})$$
$$\hat{a}_2 = 2^{-1/2}(\hat{a} + \alpha_{LO})$$

The $\hat{a}$ denotes the annihilation operator of the signal and alpha the complex amplitude of the local oscillator. The number of photon difference is eventually proportional to the quadrature and given by:

$$\hat{n}_{21} = \hat{n}_2 - \hat{n}_1 = \alpha_{LO}^* \hat{a} + \alpha_{LO} \hat{a}^\dagger$$

Rewriting this with the relation:

$$\hat{q} = 2^{-1/2}(\hat{a}^\dagger + \hat{a})$$

Results in the following relation:

$$\hat{n}_{21} = 2^{1/2} \,|\, \hat{\alpha}_{LO} \,|\, \hat{q}_\theta$$

where we see clear relation between the photon number difference and the quadrature component $\hat{q}_\theta$. By keeping track of the sum current, one can recover information about the local oscillator's intensity, since this is usually an unknown quantity, but an important quantity for calculating the quadrature component $\hat{q}_\theta$.

Problems with Linear Inversion

One of the primary problems with using linear inversion to solve for the density matrix is that in general the computed solution will not be a valid density matrix. For example, it could give negative probabilities or probabilities greater than 1 to certain measurement outcomes. This is particularly an issue when fewer measurements are made.

Another issue is that in infinite dimensional Hilbert spaces, an infinite number of measurement outcomes would be required. Making assumptions about the structure and using a finite measurement basis leads to artifacts in the phase space density.

Maximum Likelihood Estimation

Maximum likelihood estimation (also known as MLE or MaxLik) is a popular technique for dealing with the problems of linear inversion. By restricting the domain of density matrices to the proper space, and searching for the density matrix which maximizes the likelihood of giving the experimental results, it guarantees the state to be theoretically valid while giving a close fit to the data. The likelihood of a state is the probability that would be assigned to the observed results had the system been in that state.

Suppose the measurements $\{|y_j\rangle\langle y_j|\}$ have been observed with frequencies f_j. Then the likelihood associated with a state $\hat{\rho}$ is

$$L(\hat{\rho}) = \prod_j \langle y_j | \hat{\rho} | y_j \rangle^{f_j}$$

Finding the maximum of this function is non-trivial and generally involves iterative methods. The methods are an active topic of research.

Problems with Maximum Likelihood Estimation

Maximum likelihood estimation suffers from some less obvious problems than linear inversion. One problem is that it makes predictions about probabilities that cannot be justified by the data. This is seen most easily by looking at the problem of zero eigenvalues. The computed solution using MLE often contains eigenvalues which are 0, i.e. it is rank deficient. In these cases, the solution then lies on the boundary of the n-dimensional Bloch sphere. This can be seen as related to linear inversion giving states which lie outside the valid space (the Bloch sphere). MLE in these cases picks a nearby point that is valid, and the nearest points are generally on the boundary.

Bayesian Methods

Bayesian mean estimation (BME) is a relatively new approach which addresses the problems of maximum likelihood estimation. It focuses on finding optimal solutions which are also *honest* in that they include error bars in the estimate. The general idea is to start with a likelihood function and a function describing the experimenter's prior knowledge (which might be a constant function), then integrate over all density matrices using the product of the likelihood function and prior knowledge function as a weight.

Given a reasonable prior knowledge function, BME will yield a state strictly within the n-dimensional Bloch sphere. In the case of a coin flipped N times to get N heads, with a constant prior knowledge function, BME would assign $\frac{1}{N+2}$ as the probability for tails.

BME provides a high degree of accuracy in that it minimizes the *operational divergences* of the estimate from the actual state.

Methods for Incomplete Data

The number of measurements needed for a full quantum state tomography for a multi-particle system scales exponentially with the number of particles, which makes such a procedure impossible even for modest system sizes. Hence, several methods have been developed to realize quantum tomography with fewer measurements.

The concept of matrix completion and compressed sensing have been applied to reconstruct density matrices from an incomplete set of measurements (that is, a set of measurements which is not a quorum). In general, this is impossible, but under assumptions (for example, if the density matrix is a pure state, or a combination of just a few pure states) then the density matrix has fewer degrees of freedom, and it may be possible to reconstruct the state from the incomplete measurements.

Permutationally Invariant Quantum Tomography is a procedure that has been developed mostly for states that are close to being permutationally symmetric, which is typical in nowadays experiments. For two-state particles, the number of measurements needed scales only quadratically with the number of particles. Besides the modest measurement effort, the processing of the measured data can also be done efficiently: It is possible to carry out the fitting of a physical density matrix on the measured data even for large systems. Permutationally Invariant Quantum Tomography has been combined with compressed sensing in a six-qubit photonic experiment.

4.3.3 Quantum Measurement Tomography

One can imagine a situation in which an apparatus performs some measurement on quantum systems, and determining what particular measurement is desired. The strategy is to send in systems of various known states, and use these states to estimate the outcomes of the unknown measurement. Also known as "quantum estimation", tomography techniques are increasingly important including those for quantum measurement tomography and the very similar quantum state tomography. Since a measurement can always be characterized by a set of POVM's, the goal is to reconstruct the characterizing POVM's Π_l. The simplest approach is linear inversion. As in quantum state observation, use

$$\mathrm{Tr}[\Pi_l \rho_m] = \mathrm{P}(l \mid \rho_m)$$

Not surprisingly, this suffers from the same pitfalls as in quantum state tomography: namely, non-physical results, in particular negative probabilities. Here the Π_l will not be valid POVM's, as they will not be positive. Bayesian methods as well as Maximum likelihood estimation of the density matrix can be used to restrict the operators to valid physical results.

4.3.4 Quantum Process Tomography

Quantum process tomography (QPT) deals with identifying an unknown quantum dynamical process. The first approach, introduced in 1996 and sometimes known as standard quantum process tomography (SQPT) involves preparing an ensemble of quantum states and sending them through the process, then using quantum state tomography to identify the resultant states. Other techniques include ancilla-assisted process tomography (AAPT) and entanglement-assisted process tomography (EAPT) which require an extra copy of the system.

The number of experimental configurations (state preparations and measurements) required for full quantum process tomography grows exponentially with the number of constituent particles of a system. Consequently, in general, QPT is an impossible task for large-scale quantum systems. However, under weak decoherence assumption, a quantum dynamical map can find a sparse representation. The method of compressed quantum process tomography (CQPT) uses the compressed sensing technique and applies the sparsity assumption to reconstruct a quantum dynamical map from an incomplete set of measurements or test state preparations.

4.3.5 Quantum Process Tomography and Its Role in Sensing

Closely related to state tomography is the concept of quantum process tomography, which aims to reconstruct the dynamical transformation that a quantum state undergoes during its evolution or interaction with an environment. In sensing applications, this becomes relevant when the quantum sensor is exposed to an unknown field or environment.

By performing process tomography, one can understand how the external influence alters the sensor's quantum state, thereby allowing for precise inference about the external parameter. For instance, in a gravimeter based on matter-wave interferometry, process tomography can be used to map how gravitational gradients affect the evolution of the superposed atomic states. This method thus serves not only for calibration but also as a sensing tool in itself.

4.3.6 Error Analysis and Fidelity Assessment

An important part of quantum state tomography in sensing is the assessment of errors and the fidelity of reconstruction. Since quantum measurements are probabilistic and susceptible to both statistical and systematic errors, the reconstructed state might not be perfectly accurate. Quantitative measures such as trace distance, fidelity, and purity are employed to evaluate how close the reconstructed state is to the expected theoretical state. In practical sensing applications, this feedback allows engineers to tune the experimental conditions, improve isolation from environmental noise, and refine control protocols. High-fidelity state tomography assures that the sensor is accurately capturing the signal of interest, which is essential when the measurement outcomes are used to inform critical decisions, such as in navigation systems or fundamental physics experiments.

4.3.7 Tomography in Entangled and Multi-Particle Sensing

Quantum state tomography becomes particularly significant in sensors that use entanglement or collective quantum behavior. Multi-particle entangled states like GHZ or W states require sophisticated tomography schemes, as the Hilbert space scales exponentially with the number of particles. In distributed sensing networks, where entangled sensors are deployed over large distances, tomography can verify whether the entanglement is maintained and whether the correlations are consistent with theoretical predictions. This capability is crucial for emerging technologies such as quantum-enhanced radar, quantum gravimetry, and space-based quantum sensors, where the performance relies heavily on maintaining fragile quantum correlations over time and space.

4.3.8 Challenges and Future Directions

Despite its utility, quantum state tomography faces several challenges, particularly in terms of scalability, speed, and noise tolerance. Traditional full-state tomography becomes infeasible for systems beyond a few qubits or modes. There is ongoing research to develop machine learning-based methods that can infer quantum states more efficiently and adaptively. Neural networks, for instance, are being trained to predict the underlying state from partial or noisy measurement data. Another promising direction is the use of self-learning sensors that incorporate adaptive tomography in their feedback loop, thus optimizing performance dynamically in changing environments. As quantum sensors are deployed in more complex and variable conditions, such intelligent tomography methods will become indispensable.

4.4 QUANTUM PHASE ESTIMATION TECHNIQUES

Quantum Phase Estimation (QPE) is one of the most significant algorithms in quantum computing and sensing, serving as a foundational technique for extracting phase information associated with eigenvalues of unitary operators. At its core, QPE allows for the determination of the phase θ in an eigenvalue expression of the form e^(2πiθ), where θ lies between 0 and 1. This ability to precisely estimate phase is instrumental in quantum sensing because many physical quantities—such as time, frequency, and magnetic fields—can be encoded in the phase of a quantum state. As quantum technologies evolve, QPE not only contributes to advancements in computational tasks but also enhances the sensitivity and resolution of quantum sensors through precise measurements at the quantum level. Quantum Phase Estimation is a powerful technique that bridges quantum computing and quantum sensing, enabling high-precision measurement of phase information that underlies many physical quantities. Whether implemented through the standard algorithm or through iterative and adaptive variants, QPE provides a versatile framework for extracting detailed quantum information. Its applications in metrology, spectroscopy, timekeeping, and even quantum algorithm design make it a central pillar of quantum technology. As hardware improves and algorithms evolve, the influence of QPE in both theoretical and applied domains is set to expand, pushing the boundaries of what is measurable and knowable through quantum mechanics.

4.4.1 The Principle of Phase Estimation

The principle behind QPE is fundamentally quantum mechanical. When a quantum system evolves under a Hamiltonian, the eigenvalues of the system often take on a form that contains phase information. By preparing an initial state that is an eigenstate of a unitary operator and then subjecting it to controlled unitary operations, followed by an inverse quantum Fourier transform, the resulting quantum state encodes the phase in a measurable form. The final measurement on a quantum register yields the binary representation of the estimated phase with high precision. The estimation becomes increasingly accurate as the number of qubits used in the ancillary register increases, allowing for finer resolution in phase measurement.

4.4.2 Standard Quantum Phase Estimation Algorithm

The standard Quantum Phase Estimation algorithm uses two registers: one to hold the eigenstate and another to store the binary digits of the phase. The process begins by applying Hadamard gates to the ancillary qubits, creating a superposition. Then, controlled unitary operations apply phase shifts that depend on the eigenvalue of the input state. Afterward, an inverse quantum Fourier transform (QFT) is applied to the ancillary register, transforming the superposed amplitudes into a form where a final measurement yields the desired phase information. This approach has the advantage of exponential speed-up in phase estimation compared to classical methods, making it a powerful technique for both quantum algorithms and sensing applications.

4.4.3 Iterative and Adaptive Phase Estimation Methods

While the standard QPE algorithm provides high accuracy, it requires a large number of qubits and quantum gates, which poses challenges for near-term quantum devices. As a result, alternative approaches like iterative phase estimation (IPE) and adaptive phase estimation have been developed. IPE reduces resource requirements by using a single ancillary qubit and applying controlled-unitary gates in a sequence, adjusting the system based on previous measurements to iteratively refine the phase estimate. Adaptive phase estimation goes further by dynamically choosing the measurement basis in response to prior outcomes, optimizing information gain at each step. These methods, while slower in some contexts, are better suited to noisy intermediate-scale quantum (NISQ) devices and real-world quantum sensors.

4.4.4 Quantum Phase Estimation in Metrology

Quantum metrology deals with high-precision measurements using quantum systems, and QPE plays a crucial role in this field. Many sensing tasks involve the accumulation of a phase that depends on an external parameter—such as a magnetic field or gravitational acceleration—and the goal is to estimate this phase with minimal uncertainty. QPE enables quantum sensors to approach or even surpass the standard quantum limit, reaching the Heisenberg limit in ideal scenarios. This results in measurements that are more sensitive than any possible classical sensor. For instance, in atomic interferometry, QPE is used to estimate the phase accumulated by atoms traversing different paths under the influence of gravity or rotation, leading to high-precision gravimeters and gyroscopes.

4.4.5 Quantum Resources and Circuit Requirements

Quantum phase estimation is demanding in terms of quantum resources. The standard algorithm requires multiple qubits and precise control of entangling gates, particularly controlled-unitary gates for powers of the unitary operator. The circuit depth grows with the precision required, as higher powers of the unitary operator must be applied to capture more binary digits of the phase. Furthermore, an accurate inverse quantum Fourier transform is essential, which itself involves a complex sequence of gate operations. As such, implementing QPE on a physical quantum processor or quantum sensor requires careful engineering to minimize decoherence, gate errors, and readout noise, all of which can degrade the precision of the phase estimation.

4.4.6 Applications in Frequency and Time Measurements

One of the most practical applications of QPE lies in the measurement of frequencies and time intervals, especially in quantum clocks and spectroscopy. In these systems, phase differences that arise due to frequency mismatches can be measured using QPE to determine the frequency with high accuracy.

This principle is the basis of atomic clocks, where transitions between quantum energy levels are used to define the passage of time. By using phase estimation, one can lock a laser or oscillator to the transition frequency with extremely low uncertainty. Quantum-enhanced versions of these clocks can leverage QPE and entangled states to further reduce measurement noise and increase timekeeping precision.

4.4.7 Noise and Error Considerations

In realistic settings, noise and imperfections in the quantum system can significantly affect the outcomes of phase estimation. Decoherence, imperfect gate operations, and readout errors can lead to incorrect phase estimates. These errors are especially problematic in standard QPE due to the reliance on long coherence times and deep quantum circuits. Iterative and Bayesian versions of phase estimation help mitigate these issues by using fewer resources and incorporating error models into the estimation process. Quantum error correction and dynamical decoupling techniques can also be employed to preserve coherence during the estimation sequence, ensuring more accurate and reliable measurements even in noisy environments.

4.4.8 Advances in Quantum Algorithms Using Phase Estimation

Beyond sensing, QPE is central to several quantum algorithms that promise exponential speed-ups. Notably, it is a key subroutine in Shor's algorithm for factoring large integers, and in the Harrow-Hassidim-Lloyd (HHL) algorithm for solving linear systems. In these contexts, the algorithm uses QPE to extract eigenvalues of certain matrices, enabling solutions to computational problems that are infeasible for classical computers. The development of quantum algorithms based on QPE has indirectly benefitted quantum sensing by advancing techniques for coherent control, error mitigation, and qubit scalability—all of which are necessary for high-performance quantum sensors.

4.4.9 Future Directions in Phase Estimation for Sensing

The future of QPE in quantum sensing looks promising, especially with continued advances in quantum hardware and algorithm design. Researchers are exploring hybrid classical-quantum algorithms that combine QPE with classical optimization techniques to improve robustness and scalability. Furthermore, the integration of QPE with machine learning is being investigated to allow for faster and more adaptive phase estimation, particularly in environments with unknown or rapidly changing parameters. Miniaturization and portability of quantum sensors that use QPE will also become feasible as qubit coherence times and control fidelity improve. Ultimately, QPE will remain a cornerstone of quantum-enhanced measurement techniques, driving innovations in navigation, communication, geophysics, and fundamental science.

4.5 APPLICATIONS OF SQUEEZED STATES IN QUANTUM SENSING

Squeezed states represent a remarkable feature of quantum systems, where the uncertainty in one observable is reduced below the standard quantum limit at the expense of increased uncertainty in its conjugate variable. Rooted in the principles of the Heisenberg uncertainty relation, squeezed states exploit the inherent flexibility of quantum fluctuations to achieve enhanced precision in measurements. This quality makes squeezed states especially useful in quantum sensing, a field that aims to utilize quantum properties for high-accuracy measurements of physical quantities such as time, acceleration, magnetic fields, and gravitational forces. The use of squeezed states in quantum sensing is rapidly gaining momentum due to their capacity to suppress noise and surpass classical performance bounds.

The application of squeezed states in quantum sensing represents a confluence of deep theoretical insights and practical technological innovation. By harnessing quantum correlations to suppress noise and enhance sensitivity, squeezed states enable measurements that surpass classical limits in accuracy and resolution. From gravitational wave detection to atomic clocks, quantum imaging, and space exploration, the impact of squeezed states is both profound and far-reaching. As quantum technologies continue to evolve, the development of more robust, scalable, and user-friendly squeezed-state sensors will likely redefine the standards of precision in science, engineering, and industry. The future of quantum sensing, deeply intertwined with squeezed states, promises to unlock new realms of discovery and technological capability.

4.5.1 The Physics behind Squeezed States

In a standard quantum state, such as a coherent state of light, the uncertainties in position and momentum (or, analogously, amplitude and phase quadratures in optical systems) are equally distributed and constrained by the uncertainty principle. However, in a squeezed state, one of these variables exhibits a reduced variance, while the other's uncertainty correspondingly increases. This squeezing of quantum noise enables improved sensitivity to perturbations along the "squeezed" variable. Technologically, squeezed states are often produced using nonlinear optical processes, such as parametric down-conversion or four-wave mixing, where an intense pump laser interacts with a nonlinear medium to redistribute quantum noise. The result is a state of light or matter with quantum correlations that are essential for enhanced sensing capabilities.

4.5.2 Enhanced Optical Interferometry Using Squeezed Light

One of the earliest and most prominent applications of squeezed states in quantum sensing lies in optical interferometry. Classical interferometers, like the Michelson interferometer, are limited by shot noise—a quantum noise that sets the standard quantum limit on sensitivity. By injecting squeezed light into the unused port of an interferometer, the shot noise can be reduced, allowing for enhanced phase sensitivity.

This technique has been applied in gravitational wave observatories such as LIGO and Virgo, where squeezed light improves the detectors' ability to observe minute spacetime distortions caused by distant astrophysical events. The successful integration of squeezed states in these observatories has marked a milestone in both fundamental physics and applied quantum sensing.

4.5.3 Squeezed States in Atomic Sensors and Clocks

Atomic systems also benefit greatly from squeezed states. In atomic clocks and magnetometers, quantum noise in the population of atoms or their collective spin can limit measurement precision. By preparing atoms in spin-squeezed states, the variance in the population difference or spin projection is reduced, leading to more precise estimation of frequency or magnetic field strength. This approach has enabled the development of atomic sensors with performance exceeding the standard quantum limit. Spin squeezing is typically achieved through quantum non-demolition measurements or via interactions mediated by optical cavities, both of which introduce correlations among atoms that redistribute quantum uncertainty. These techniques open up possibilities for ultraprecise navigation, timing, and geophysical surveys.

4.5.4 Quantum Metrology and the Heisenberg Limit

The fundamental limit of measurement precision in quantum systems is given by the Heisenberg limit, which scales inversely with the total number of particles or resources used. Squeezed states enable quantum sensors to approach or reach this limit by minimizing the uncertainty in the measurement variable of interest.

In quantum metrology, this capability is particularly valuable when dealing with systems where resource consumption must be minimized or where maximizing precision is essential. Squeezed states thus provide a pathway to surpass classical limitations and are integral to the next generation of quantum-enhanced metrological standards and instruments.

4.5.5 Sensing in Noisy Environments with Squeezed States

In real-world sensing applications, environmental noise and decoherence can degrade measurement accuracy. Interestingly, squeezed states are not only useful for reducing intrinsic quantum noise but also for making measurements more robust against certain types of external disturbances. For example, squeezing can enhance the signal-to-noise ratio in systems where classical noise sources dominate along specific quadratures. In optical systems, the ability to tailor the squeezing direction allows for dynamic adaptation to noise characteristics, thus optimizing measurement fidelity even in fluctuating environments. This adaptability is especially useful in fields such as biomedical imaging, where detecting faint signals amidst significant background noise is often required.

4.5.6 Quantum Imaging and Spectroscopy

The precision benefits of squeezed states extend beyond scalar measurements into imaging and spectroscopy. In quantum imaging, squeezed light can be used to illuminate a sample in such a way that the quantum correlations reduce noise and increase resolution, even below the diffraction limit in some cases. This has implications for microscopy techniques aimed at observing delicate biological systems without excessive exposure to damaging light levels. In spectroscopy, squeezed states can enable more accurate measurement of spectral lines by reducing the uncertainty in frequency measurements. These capabilities are pushing the boundaries of how we observe and analyze microscopic systems, both in material science and life sciences.

4.5.7 Quantum Magnetometry with Squeezed Ensembles

Magnetometry, the measurement of magnetic fields, is another domain where squeezed states offer remarkable advantages. Quantum magnetometers using ensembles of atoms or NV centers in diamond can achieve higher sensitivity when the constituent particles are prepared in squeezed spin states. These devices can detect extremely weak magnetic fields, such as those produced by neuronal activity in the brain or geological anomalies underground. The use of squeezed states amplifies the magnetic field signal without increasing susceptibility to noise along orthogonal axes. This makes squeezed-state quantum magnetometers highly promising for non-invasive medical diagnostics, geological exploration, and even underwater navigation.

4.5.8 Quantum Sensors in Space and Fundamental Physics

Squeezed states are also expected to play an important role in space-based quantum sensing missions. In microgravity environments, where classical sensors can be challenged by noise and instability, quantum sensors based on squeezed light or atoms provide enhanced accuracy for applications such as gravitational field mapping or satellite navigation. Moreover, squeezed states are increasingly being considered for fundamental physics experiments that test the limits of general relativity or search for dark matter. In such experiments, detecting subtle deviations from known physics requires exceptional measurement precision, which squeezed states can deliver by enhancing interferometric sensitivity beyond the reach of classical techniques.

4.5.9 Challenges and Technological Developments

Despite their many advantages, practical deployment of squeezed states in sensing systems comes with challenges. Generating and maintaining squeezed states over long durations and in diverse conditions remains technologically demanding. Losses in optical components, decoherence in atomic systems, and the complexity of integrating squeezed state sources with conventional sensor platforms all pose significant hurdles. Nonetheless, advances in photonic engineering, cryogenics, and quantum control techniques are steadily addressing these issues. Compact sources of squeezed light are being developed using integrated photonics, and new materials and protocols are extending coherence times for squeezed atomic states. As these technologies mature, the scope of applications for squeezed-state-based sensors is expected to broaden significantly.

4.6 COHERENT POPULATION TRAPPING AND ITS APPLICATIONS IN SENSING

Coherent Population Trapping (CPT) is a quantum optical phenomenon in which atoms or molecules are driven into a non-absorbing superposition state, known as a "dark state," by the application of two coherent laser fields. This state results from quantum interference between different excitation pathways, effectively preventing the system from absorbing light. First observed in the late 1970s, CPT has since become a powerful tool in quantum optics, atomic physics, and metrology. Its importance lies in the ability to manipulate and probe atomic coherence with high precision, making it especially relevant for applications in quantum sensing, where sensitivity and accuracy are paramount. Coherent Population Trapping is a foundational quantum optical phenomenon that has found diverse and impactful applications in sensing.

By leveraging quantum interference to isolate systems in non-absorbing coherent states, CPT provides a pathway to achieve ultra-sensitive, precise, and compact measurement devices. Its contributions span a range of disciplines, including timekeeping, magnetometry, environmental monitoring, biomedical diagnostics, and quantum information science. As the field of quantum sensing continues to evolve, the role of CPT is set to grow in prominence, supported by advancements in laser technology, atomic physics, and device engineering. The journey from fundamental physics to practical application highlights the transformative power of quantum principles in solving real-world challenges and enabling new technological frontiers.

4.6.1 The Physics behind Coherent Population Trapping

CPT occurs in a three-level atomic system typically arranged in a Λ-type configuration, where two lower energy states are coupled to an excited state via two laser fields. When the frequency difference of the two lasers matches the energy difference between the two ground states, destructive interference between the excitation pathways can eliminate absorption. The system then becomes trapped in a coherent superposition of the two ground states that does not interact with the light, hence the term "dark state." This suppression of absorption is highly sensitive to the relative frequency and phase of the applied fields, which is the key to its utility in precision measurements.

4.6.2 Atomic Clocks and Frequency Standards

One of the most significant applications of CPT is in the development of compact and highly accurate atomic clocks. Traditional atomic clocks require microwave cavities and complex setups to probe atomic transitions. CPT-based atomic clocks, on the other hand, use laser fields to induce coherent superpositions in alkali atoms such as rubidium or cesium. The dark state formed by CPT has a narrow spectral linewidth and is highly stable against fluctuations, making it an excellent reference for frequency standards. These clocks are more compact, consume less power, and are particularly suitable for deployment in mobile and space-based systems where conventional atomic clocks are impractical.

4.6.3 Magnetometry Using CPT

CPT has also been widely adopted in the field of magnetometry, where the aim is to measure magnetic fields with extreme precision. In CPT-based magnetometers, the frequency of the dark resonance shifts in the presence of a magnetic field due to the Zeeman effect. By monitoring this shift, it is possible to deduce the strength and orientation of the magnetic field. These devices are capable of detecting extremely weak fields, down to femtotesla levels, making them suitable for applications ranging from biomedical imaging of brain activity to detecting unexploded ordnance and mapping geophysical anomalies. The non-invasive nature and room-temperature operation of CPT magnetometers make them highly attractive in diverse sensing scenarios.

4.6.4 Temperature and Pressure Sensing

CPT's sensitivity to environmental parameters such as temperature and pressure allows it to be adapted for other forms of sensing. The coherence properties of the atomic medium, and hence the CPT signal, can be influenced by collisions between atoms, which are in turn affected by temperature and pressure. This relationship can be harnessed to construct sensors capable of monitoring these parameters with high resolution. Although such applications are still in the developmental stage compared to magnetometry or atomic clocks, they hold promise for compact, non-invasive environmental sensors that could be integrated into portable or embedded systems.

4.6.5 CPT in Quantum Information and Control

Beyond classical sensing applications, CPT plays a vital role in quantum information science. It is instrumental in achieving long-lived quantum coherence, which is essential for quantum memory and quantum computing. In the context of sensing, this coherence allows for more robust quantum state manipulation and longer interrogation times, which translate into greater measurement precision. CPT-based techniques can help initialize and read out quantum states in a non-destructive manner, thus preserving the quantum system for repeated measurements. This attribute is particularly valuable in the development of quantum networks and distributed sensing systems.

4.6.6 Advances in CPT-Based Laser Cooling

Another intriguing application of CPT is in laser cooling, where it has been used to reach temperatures lower than the Doppler cooling limit. The formation of dark states reduces photon scattering, thereby limiting the heating effects typically associated with laser-atom interactions. Such cooling techniques enhance the performance of atomic sensors by reducing thermal motion, which otherwise contributes to noise and decoherence. This advance is particularly relevant for precision sensors used in navigation, inertial guidance, and gravitational studies, where ultracold atoms provide superior signal stability and resolution.

4.6.7 Miniaturization and Integration of CPT Sensors

A notable advantage of CPT-based systems is their potential for miniaturization and integration into chip-scale devices. The all-optical nature of CPT means that it does not require bulky microwave resonators or cryogenic environments. As a result, CPT atomic clocks and magnetometers have been successfully implemented in compact form factors, leading to the development of chip-scale atomic clocks (CSACs) and portable magnetometers. These innovations are crucial for enabling quantum sensing in everyday technologies such as smartphones, autonomous vehicles, and wearable health monitors. With further advancements in photonics and nanofabrication, the future holds promise for even more compact and efficient CPT-based devices.

4.6.8 CPT in Biomedical and Neuroscience Applications

The ability of CPT-based sensors to detect extremely weak magnetic fields opens up exciting possibilities in biomedical research. For example, magnetoencephalography (MEG), which maps brain activity by detecting magnetic fields produced by neural currents, can benefit from CPT magnetometers. These devices provide a lightweight, room-temperature alternative to traditional superconducting magnetometers, which require cryogenic cooling. Similarly, CPT-based sensors could be used in cardiology for detecting magnetic signatures of heart activity or in diagnostic tools for imaging small-scale physiological processes. The non-invasive nature of CPT sensing makes it especially appealing for continuous and wearable health monitoring.

4.6.9 Challenges and Future Directions

Despite their many advantages, CPT-based sensors face challenges that must be addressed for widespread adoption. One of the primary limitations is sensitivity to environmental noise, such as laser frequency fluctuations and temperature instability, which can affect the coherence of the atomic dark states. In addition, the need for precise control over laser parameters and atomic vapor conditions necessitates sophisticated feedback systems. Nevertheless, ongoing research is focused on improving the robustness and resilience of CPT systems through better shielding, stabilization techniques, and advanced materials. Moreover, combining CPT with other quantum techniques, such as spin squeezing and entanglement, could further enhance sensitivity and reduce noise.

CHAPTER 5

Quantum Sensing in Various Applications

5.1 QUANTUM SENSORS IN NAVIGATION AND POSITIONING

5.1.1 Quantum Accelerometers for Inertial Navigation

Quantum accelerometers represent a cutting-edge class of inertial sensors that utilize the principles of quantum mechanics to measure acceleration with unprecedented precision. These devices exploit quantum properties such as matter-wave interference, atomic coherence, and superposition to detect tiny changes in velocity or movement. Unlike traditional accelerometers that depend on mechanical or electrical feedback systems, quantum accelerometers rely on the behavior of ultracold atoms and their interference patterns when subjected to acceleration. This innovative approach has opened new avenues in navigation, geophysics, defense, and aerospace applications where precise inertial measurements are critical. Quantum accelerometers are poised to revolutionize the field of inertial navigation by providing precise, drift-free acceleration measurements based on fundamental quantum principles.

Leveraging atom interferometry and cold atom physics, these sensors outperform traditional accelerometers in stability and accuracy, especially in GPS-denied or extreme environments. Although technical and engineering challenges remain, rapid progress in quantum technology and miniaturization promises to overcome these barriers. As their capabilities become more accessible, quantum accelerometers will play a critical role in the future of navigation, exploration, and scientific discovery.

The Limitations of Classical Accelerometers

Conventional accelerometers, though widely used in devices ranging from smartphones to aircraft, face intrinsic limitations. These sensors typically work on the principle of Newton's second law, using a mass suspended on a spring or microelectromechanical systems (MEMS) to measure acceleration. While adequate for short-term guidance, they suffer from issues like signal drift, thermal noise, and cumulative error over time. In applications such as submarine or spacecraft navigation where GPS signals are unavailable, these inaccuracies can lead to significant positional errors. Consequently, the need for more stable and accurate sensors has driven interest in quantum-based alternatives that do not degrade in performance over time.

Fundamentals of Quantum Accelerometry

At the heart of a quantum accelerometer is atom interferometry—a technique that uses the wave nature of atoms to measure acceleration. In a typical setup, a cloud of atoms is cooled using laser cooling techniques to form an ultracold atomic ensemble. These atoms are then manipulated with a sequence of laser pulses that split, redirect, and recombine their quantum wavefunctions, creating an interference pattern. Any acceleration experienced by the system alters the phase of the atomic waves, which is then measured to determine the exact movement of the system. The advantage of this method is its extreme sensitivity, derived from the fundamental stability and predictability of atomic behavior.

Role in Inertial Navigation Systems

Inertial navigation systems (INS) are critical for determining the position and velocity of a vehicle without relying on external references. These systems integrate data from gyroscopes and accelerometers to compute navigation information. Quantum accelerometers greatly enhance the capabilities of INS by offering drift-free measurements over long durations. Their high stability and precision make them especially valuable in GPS-denied environments such as underwater navigation for submarines, underground exploration, and interplanetary space missions. They can operate independently of satellite signals and external communication, offering secure and robust navigation capabilities.

Advancements in Cold Atom Technology

The development of quantum accelerometers has been closely linked to advancements in cold atom physics. Techniques such as magneto-optical trapping and laser cooling allow atoms to be brought to microkelvin temperatures, reducing their thermal motion and allowing more precise interferometric measurements. Additionally, the use of Raman transitions and optical lattices has improved the control and manipulation of atomic states. These developments have led to more compact, reliable, and accurate atomic interferometers, making the transition from laboratory prototypes to field-deployable devices increasingly feasible.

Applications in Aerospace and Defense

Quantum accelerometers have significant implications for aerospace and defense sectors. In aviation, they can serve as backup navigation systems in the event of GPS signal loss due to jamming or interference. For submarines, which often operate in GPS-inaccessible environments, quantum accelerometers offer continuous and autonomous navigation without surfacing. In missile guidance systems, their high precision can ensure accurate targeting even when GPS signals are deliberately disrupted. Their robustness and resistance to electromagnetic interference make them ideal for environments where traditional sensors may fail.

Use in Earth Observation and Geophysics

Beyond navigation, quantum accelerometers have found applications in Earth science and geophysics. Their ability to detect minute variations in acceleration allows them to sense changes in gravitational fields, making them useful for mapping geological structures, monitoring tectonic activity, and detecting underground cavities. These capabilities are valuable for oil and mineral exploration, as well as environmental monitoring. For instance, by tracking subtle gravitational anomalies, researchers can detect groundwater movement, volcanic activity, or the structural integrity of large-scale infrastructure.

Miniaturization and Technological Challenges

Despite their promise, quantum accelerometers still face technical challenges before achieving widespread adoption. One major hurdle is miniaturization. The complexity of cold atom systems, including vacuum chambers, lasers, and magnetic shielding, has traditionally required laboratory-scale setups. However, recent progress in photonics, microfabrication, and integration technologies is enabling the development of more compact and portable quantum sensors. Power consumption, thermal management, and vibration isolation also pose challenges for deployment in dynamic environments like aircraft or vehicles. Addressing these challenges is the focus of ongoing research and engineering efforts.

Comparing Quantum and Classical Inertial Sensors

When compared to their classical counterparts, quantum accelerometers offer significant advantages in long-term accuracy and stability. While MEMS-based accelerometers are small and inexpensive, they accumulate error quickly over time. In contrast, quantum accelerometers exhibit negligible drift, providing more reliable readings over extended periods. However, classical sensors still hold an advantage in terms of cost, size, and ease of integration. Therefore, hybrid systems that combine classical and quantum sensors are emerging as a practical solution, using quantum measurements for periodic recalibration of conventional sensors.

Future Outlook and Research Directions

The future of quantum accelerometry is closely tied to advances in quantum technology and the broader field of quantum sensing. Research is focused on developing all-optical systems that eliminate the need for bulky magnetic fields, increasing the interrogation time of atomic interferometers, and enhancing signal-to-noise ratios. Efforts are also being made to improve environmental robustness, enabling quantum accelerometers to function in a wider range of conditions. As national and international investments in quantum technology continue to grow, we can expect to see quantum accelerometers integrated into commercial navigation systems, autonomous vehicles, and next-generation spacecraft.

5.1.2 Quantum Gyroscopes and their Role in GPS-Free Navigation

Quantum gyroscopes are advanced navigation instruments that utilize the principles of quantum mechanics to detect rotation with extreme precision. Unlike traditional mechanical or optical gyroscopes, quantum gyroscopes exploit quantum phenomena such as matter-wave interference, spin precession, and atomic coherence.

This allows them to measure angular velocity with far greater stability and sensitivity. As the demand for highly accurate, autonomous navigation systems grows—especially in environments where GPS signals are unavailable or unreliable—quantum gyroscopes have emerged as vital tools in aerospace, defense, marine, and underground applications. Their development represents a major step forward in the quest for GPS-free navigation solutions. Quantum gyroscopes represent a transformative leap in navigation technology, offering a powerful solution to the limitations of GPS-dependent systems.

By harnessing quantum principles such as atom interferometry and matter-wave interference, these devices provide exceptionally precise and stable measurements of angular velocity. As engineering challenges related to miniaturization and robustness are overcome, quantum gyroscopes are poised to revolutionize navigation in both military and civilian sectors. In an increasingly interconnected world where accurate positioning is critical and GPS vulnerabilities are a growing concern, quantum gyroscopes stand as a beacon of reliable, autonomous navigation for the future.

Limitations of Conventional Gyroscopes

Traditional gyroscopes, whether mechanical, optical, or MEMS-based (Micro-Electro-Mechanical Systems), have long been used in navigation systems to determine orientation and angular velocity. However, they suffer from several limitations, including signal drift over time, sensitivity to environmental disturbances, and limited long-term accuracy. Fiber-optic and ring-laser gyroscopes improved on earlier designs by using the interference of light to detect rotation, but they are still constrained by optical noise, thermal fluctuations, and calibration requirements. As these errors accumulate, the positional accuracy of systems relying on conventional gyroscopes diminishes over time, making them less reliable for long-duration, autonomous operations.

The Quantum Advantage in Rotation Sensing

Quantum gyroscopes overcome many of the limitations of traditional devices by leveraging the quantum nature of atoms. The most common form, known as an atom interferometric gyroscope, uses ultracold atoms manipulated by lasers to form atomic wave packets. These packets are split and made to follow different paths before being recombined. If the system rotates during this process, a measurable phase shift occurs in the resulting interference pattern. This phase shift directly corresponds to the rate of rotation, allowing the quantum gyroscope to determine angular velocity with incredible accuracy. Because atoms are far less susceptible to environmental noise than light, quantum gyroscopes offer greater long-term stability and reduced drift.

Principles of Operation in Quantum Gyroscopes

The core technology behind quantum gyroscopes is atom interferometry, which draws on the dual nature of atoms as both particles and waves. When a group of atoms is cooled to near absolute zero using laser cooling and magnetic trapping techniques, their thermal motion is drastically reduced. These atoms are then subjected to a sequence of laser pulses that effectively split and redirect their wavefunctions, creating separate paths in space. During rotation, the Sagnac effect causes a relative phase shift between the two paths. This phase difference is detected through changes in the interference pattern when the atom clouds are recombined. The amount of phase shift is directly proportional to the angular rotation experienced during the interferometric cycle.

Applications in GPS-Denied Environments

One of the most compelling uses of quantum gyroscopes is in GPS-free navigation. Many critical operations occur in environments where GPS signals are blocked, jammed, or degraded, such as deep underwater, inside tunnels or buildings, in space, or in military operations where GPS denial is deliberate. Quantum gyroscopes can maintain precise orientation and positional information over long durations without the need for satellite updates or external calibration. This makes them essential components in submarines, spacecraft, autonomous vehicles, and defense systems where reliance on GPS is either impractical or a security risk.

Integration into Inertial Navigation Systems

Inertial navigation systems (INS) combine accelerometers and gyroscopes to calculate the position, orientation, and velocity of a moving object. The accuracy of an INS is fundamentally limited by the precision of its component sensors. By replacing conventional gyroscopes with quantum gyroscopes, the performance of these systems can be dramatically enhanced. Quantum gyroscopes can provide angular velocity measurements with far less drift, allowing the INS to maintain accurate navigation for extended periods. This is particularly valuable in long-endurance missions, such as intercontinental missile guidance, planetary exploration, and unmanned underwater vehicle (UUV) operations.

Technological Challenges and Development

While quantum gyroscopes offer many advantages, their widespread deployment faces technical and engineering challenges. The systems required for atom interferometry are complex, typically involving vacuum chambers, ultra-stable lasers, magnetic field control, and vibration isolation. These setups are often large and delicate, limiting their practical use outside of laboratory settings. However, significant progress is being made toward miniaturization and ruggedization of quantum sensing equipment. Advances in chip-scale atom traps, fiber-coupled laser systems, and robust control electronics are making it possible to envision field-deployable quantum gyroscopes that can function reliably in real-world conditions.

Comparison with Optical Gyroscopes

Optical gyroscopes, such as ring laser gyroscopes and fiber optic gyroscopes, have been the mainstay of high-performance navigation for decades. They use light beams traveling in opposite directions in a closed loop, and measure the interference pattern resulting from the Sagnac effect. While highly effective, optical gyroscopes are still limited by factors such as photon shot noise and thermal effects. Quantum gyroscopes, using matter waves rather than light, can achieve a higher sensitivity because atoms have much shorter wavelengths than photons. Additionally, quantum sensors are inherently immune to many forms of electromagnetic interference, making them more robust in hostile environments.

Defense and Strategic Implications

Quantum gyroscopes have strategic importance in defense and national security. Military platforms that rely heavily on GPS navigation—such as fighter jets, tanks, and missiles—are vulnerable to signal jamming or spoofing. A quantum gyroscope-based INS offers a self-contained, highly reliable alternative that ensures operational continuity even under GPS-denied conditions. Nations investing in quantum navigation technologies are likely to gain a strategic edge in terms of autonomous mobility, precision strike capability, and secure operations. Consequently, quantum gyroscopes are receiving significant attention in defense research programs around the world.

Future Prospects and Commercial Applications

As quantum technologies continue to evolve, the potential for commercial applications of quantum gyroscopes is rapidly expanding. Future aircraft, spacecraft, and even autonomous cars may benefit from the precise, drift-free orientation provided by quantum sensors. Oil and gas exploration, high-speed rail systems, and civil aviation can also utilize quantum gyroscopes for more accurate motion tracking and orientation. In addition, coupling quantum gyroscopes with other quantum sensors, such as quantum accelerometers and magnetometers, can create integrated quantum navigation units capable of operating entirely independently of external signals.

5.2 QUANTUM SENSORS IN GRAVITATIONAL WAVE DETECTION

5.2.1 Applications in LIGO and Future Gravitational Wave Observatories

Gravitational waves are ripples in spacetime caused by the acceleration of massive objects, such as the merging of black holes or neutron stars. These waves travel at the speed of light and carry information about their origins, offering a novel way of studying some of the most extreme and enigmatic phenomena in the universe. The Laser Interferometer Gravitational-Wave Observatory (LIGO) is one of the key instruments designed to detect these elusive waves. LIGO's success in detecting gravitational waves has opened up an entirely new field of astronomy, often referred to as "gravitational wave astronomy." The detection of gravitational waves by LIGO in 2015 marked a historic milestone, providing a new tool for exploring the universe. LIGO's groundbreaking detections of gravitational waves have revolutionized our understanding of the universe. The ability to directly observe cosmic events through gravitational waves has provided insights into the fundamental forces of nature, tested general relativity, and opened up new avenues for studying black holes, neutron stars, and other extreme objects. As the technology behind gravitational wave observatories continues to advance, future observatories such as LISA and the Einstein Telescope promise to deepen our understanding of the cosmos even further.

The role of gravitational wave observatories in exploring the universe's most enigmatic phenomena will continue to expand, offering new discoveries that could change our understanding of space, time, and gravity. As we look to the future, the applications of gravitational wave technology will also likely extend into areas such as Earth sciences and fundamental physics, offering insights into our own planet and the universe's deepest mysteries.

The Role of LIGO in Gravitational Wave Detection

LIGO is based on a technique called laser interferometry, which is designed to detect extremely minute changes in distance caused by passing gravitational waves. The LIGO facility consists of two observatories, one in Livingston, Louisiana, and the other in Hanford, Washington. These observatories are equipped with large laser interferometers that measure the distance between mirrors placed kilometers apart. When a gravitational wave passes through the Earth, it stretches and compresses spacetime, which results in tiny changes in the distance between the mirrors. LIGO's lasers detect these minute fluctuations, allowing scientists to measure the passing waves. The detection of gravitational waves requires extreme precision. The changes in distance caused by gravitational waves are on the order of one-thousandth the diameter of a proton, a level of precision that challenges even the most advanced scientific equipment. However, LIGO's sensitivity and technological innovations allow it to detect these faint signals, often from events that occurred billions of years ago. These detections have not only confirmed key aspects of Einstein's theory of general relativity but also provided insights into the behavior of matter and energy under extreme conditions.

Applications of LIGO in Astrophysics

One of the primary applications of LIGO is in the study of astrophysical phenomena. Gravitational waves provide a new way to observe and study objects and events in the universe that are otherwise invisible through traditional electromagnetic observations, such as light or radio waves. For instance, the detection of gravitational waves from the collision of black holes or neutron stars has provided astronomers with new insights into the nature of these objects. Before the detection of gravitational waves, scientists could only infer the existence of black holes from their interactions with nearby matter or through their effect on the surrounding space. However, gravitational waves offer a direct method of observing black hole mergers and the dynamics of these extreme systems. The first detection of gravitational waves in 2015, known as GW150914, came from the merger of two black holes, providing direct evidence of their existence and confirming many theoretical predictions about their behavior. LIGO has also played a crucial role in the study of neutron stars. The detection of gravitational waves from the merger of two neutron stars, which occurred in 2017 (GW170817), provided valuable insights into the formation of heavy elements like gold and platinum. The accompanying electromagnetic signal—detected by telescopes around the world—allowed scientists to observe the merger in both gravitational waves and electromagnetic radiation for the first time. This multi-messenger astronomy approach has opened up new avenues for studying astrophysical events.

LIGO's Impact on Fundamental Physics

Beyond its applications in astrophysics, LIGO has significant implications for fundamental physics. The detection of gravitational waves serves as a direct test of Einstein's theory of general relativity. The theory predicts that massive objects accelerating through space-time would generate ripples in spacetime, and LIGO's detections have confirmed this prediction. This validation of general relativity has far-reaching implications for our understanding of the fabric of spacetime and the fundamental forces of nature. LIGO also plays a role in testing the limits of general relativity and probing the nature of gravity in extreme environments. For instance, the precise measurements made by LIGO may reveal new physics beyond general relativity, such as deviations from Newtonian gravity or new insights into the behavior of gravity at quantum scales. Gravitational wave astronomy opens up possibilities for exploring the universe's fundamental laws, which could lead to groundbreaking discoveries about the nature of gravity, spacetime, and the universe itself.

Challenges and Future Directions of Gravitational Wave Detection

While LIGO has achieved incredible success, the field of gravitational wave astronomy is still in its early stages. One of the main challenges is improving the sensitivity of current detectors. Gravitational waves are incredibly weak and often difficult to detect, especially when they originate from distant sources. Current detectors, including LIGO, are only sensitive enough to detect waves from the most extreme and energetic events, such as the mergers of black holes and neutron stars. For more subtle or distant events, the sensitivity of the detectors must be further improved. To address these challenges, several initiatives are underway to enhance the sensitivity and capability of existing gravitational wave observatories. LIGO is undergoing an upgrade to improve its sensitivity by a factor of 10, which will increase the range at which it can detect gravitational waves. This next phase of LIGO, known as the Advanced LIGO, is expected to detect a greater variety of events and more distant sources of gravitational waves.

Future Gravitational Wave Observatories

The success of LIGO has spurred the development of future gravitational wave observatories, both on Earth and in space. One such project is the Virgo interferometer, located in Italy, which works in collaboration with LIGO to improve the accuracy of gravitational wave detections by triangulating the source's location. The addition of the Virgo detector enhances the ability of researchers to pinpoint the location of gravitational wave events and study them in more detail. In the coming years, more advanced observatories are being planned to further expand our ability to detect gravitational waves. One of the most ambitious projects is the Laser Interferometer Space Antenna (LISA), a space-based gravitational wave observatory. LISA aims to detect low-frequency gravitational waves that originate from sources such as supermassive black hole mergers and galactic nuclei. Because LISA will be in space, it will be free from the seismic and environmental noise that limits ground-based detectors like LIGO, allowing it to detect a broader range of gravitational waves. Another future observatory is the Einstein Telescope, which is a proposed underground gravitational wave detector that will offer even greater sensitivity than LIGO. By being located underground, the Einstein Telescope will be less affected by noise from environmental factors such as seismic activity, making it better suited to detect weak gravitational wave signals from more distant or faint sources.

Applications beyond Astrophysics

As the sensitivity of gravitational wave observatories increases, it is expected that gravitational wave detection will have applications beyond traditional astrophysical studies. For example, gravitational wave sensors could potentially be used to monitor seismic activity or detect underground nuclear explosions. Additionally, as quantum technologies continue to advance, gravitational wave detectors themselves could provide valuable tools for testing new theories of quantum mechanics and understanding the nature of space and time at fundamental levels.

5.3 QUANTUM SENSING IN MEDICAL IMAGING AND DIAGNOSTICS

5.3.1 Magnetic Resonance Imaging (MRI) Enhancement with Quantum Sensors

The Basics of MRI Technology

Magnetic resonance imaging (MRI) is a medical imaging technique used in radiology to generate pictures of the anatomy and the physiological processes inside the body. MRI scanners use strong magnetic fields, magnetic field gradients, and radio waves to form images of the organs in the body. MRI does not involve X-rays or the use of ionizing radiation, which distinguishes it from computed tomography (CT) and positron emission tomography (PET) scans. MRI is a medical application of nuclear magnetic resonance (NMR) which can also be used for imaging in other NMR applications, such as NMR spectroscopy.

MRI is widely used in hospitals and clinics for medical diagnosis, staging and follow-up of disease. Compared to CT, MRI provides better contrast in images of soft tissues, e.g. in the brain or abdomen. However, it may be perceived as less comfortable by patients, due to the usually longer and louder measurements with the subject in a long, confining tube, although "open" MRI designs mostly relieve this. Additionally, implants and other non-removable metal in the body can pose a risk and may exclude some patients from undergoing an MRI examination safely.

MRI was originally called NMRI (nuclear magnetic resonance imaging), but "nuclear" was dropped to avoid negative associations. Certain atomic nuclei are able to absorb radio frequency (RF) energy when placed in an external magnetic field; the resultant evolving spin polarization can induce an RF signal in a radio frequency coil and thereby be detected. In other words, the nuclear magnetic spin of protons in the hydrogen nuclei resonates with the RF incident waves and emit coherent radiation with compact direction, energy (frequency) and phase. This coherent amplified radiation is easily detected by RF antennas close to the subject being examined. It is a process similar to masers. In clinical and research MRI, hydrogen atoms are most often used to generate a macroscopic polarized radiation that is detected by the antennas. Hydrogen atoms are naturally abundant in humans and other biological organisms, particularly in water and fat. For this reason, most MRI scans essentially map the location of water and fat in the body. Pulses of radio waves excite the nuclear spin energy transition, and magnetic field gradients localize the polarization in space. By varying the parameters of the pulse sequence, different contrasts may be generated between tissues based on the relaxation properties of the hydrogen atoms therein.

Since its development in the 1970s and 1980s, MRI has proven to be a versatile imaging technique. While MRI is most prominently used in diagnostic medicine and biomedical research, it also may be used to form images of non-living objects, such as mummies. Diffusion MRI and functional MRI extend the utility of MRI to capture neuronal tracts and blood flow respectively in the nervous system, in addition to detailed spatial images. The sustained increase in demand for MRI within health systems has led to concerns about cost effectiveness and overdiagnosis.

Construction and Physics

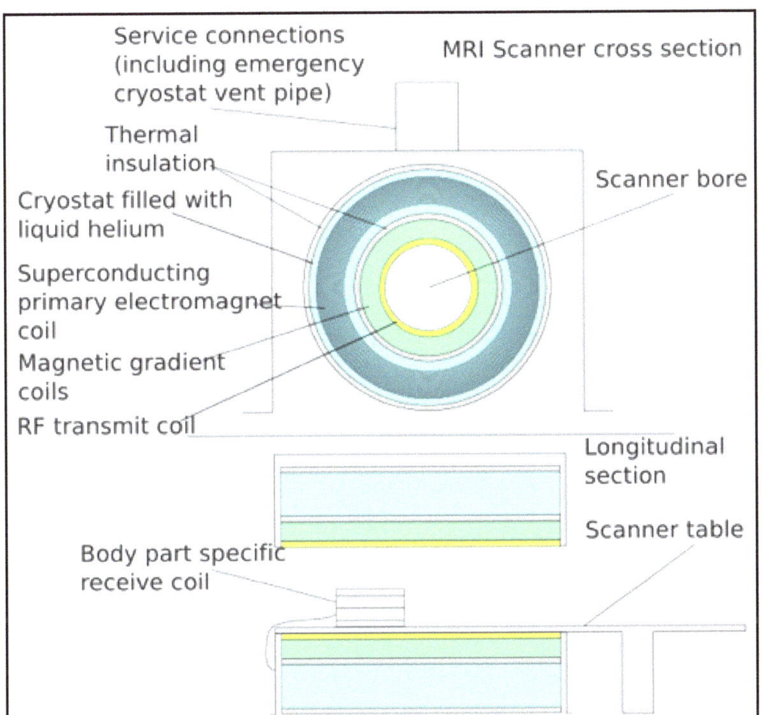

Fig: Schematic of a Cylindrical Superconducting MR Scanner

Top: cross section of the cylinder with primary coil, gradient coils and RF transmit coils. Bottom: longitudinal section of the cylinder and table, showing the same coils and the RF receive coil.

In most medical applications, hydrogen nuclei, which consist solely of a proton, that are in tissues create a signal that is processed to form an image of the body in terms of the density of those nuclei in a specific region. Given that the protons are affected by fields from other atoms to which they are bonded, it is possible to separate responses from hydrogen in specific compounds. To perform a study, the person is positioned within an MRI scanner that forms a strong magnetic field around the area to be imaged. First, energy from an oscillating magnetic field is temporarily applied to the patient at the appropriate resonance frequency. Scanning with X and Y gradient coils causes a selected region of the patient to experience the exact magnetic field required for the energy to be absorbed. The atoms are excited by a RF pulse and the resultant signal is measured by a receiving coil. The RF signal may be processed to deduce position information by looking at the changes in RF level and phase caused by varying the local magnetic field using gradient coils. As these coils are rapidly switched during the excitation and response to perform a moving line scan, they create the characteristic repetitive noise of an MRI scan as the windings move slightly due to magnetostriction. The contrast between different tissues is determined by the rate at which excited atoms return to the equilibrium state. Exogenous contrast agents may be given to the person to make the image clearer.

The major components of an MRI scanner are the main magnet, which polarizes the sample, the shim coils for correcting shifts in the homogeneity of the main magnetic field, the gradient system which is used to localize the region to be scanned and the RF system, which excites the sample and detects the resulting NMR signal. The whole system is controlled by one or more computers.

MRI requires a magnetic field that is both strong and uniform to a few parts per million across the scan volume. The field strength of the magnet is measured in teslas – and while the majority of systems operate at 1.5 T, commercial systems are available between 0.2 and 7 T. 3T MRI systems, also called 3 Tesla MRIs, have stronger magnets than 1.5 systems and are considered better for images of organs and soft tissue. Whole-body MRI systems for research applications operate in e.g. 9.4T, 10.5T, 11.7T. Even higher field whole-body MRI systems e.g. 14 T and beyond are in conceptual proposal or in engineering design. Most clinical magnets are superconducting magnets, which require liquid helium to keep them at low temperatures. Lower field strengths can be achieved with permanent magnets, which are often used in "open" MRI scanners for claustrophobic patients. Lower field strengths are also used in a portable MRI scanner approved by the FDA in

2020. Recently, MRI has been demonstrated also at ultra-low fields, i.e., in the microtesla-to-millitesla range, where sufficient signal quality is made possible by prepolarization (on the order of 10–100 mT) and by measuring the Larmor precession fields at about 100 microtesla with highly sensitive superconducting quantum interference devices (SQUIDs).

T1 and T2

Each tissue returns to its equilibrium state after excitation by the independent relaxation processes of T_1 (spin-lattice; that is, magnetization in the same direction as the static magnetic field) and T_2 (spin-spin; transverse to the static magnetic field). To create a T_1-weighted image, magnetization is allowed to recover before measuring the MR signal by changing the repetition time (TR). This image weighting is useful for assessing the cerebral cortex, identifying fatty tissue, characterizing focal liver lesions, and in general, obtaining morphological information, as well as for post-contrast imaging. To create a T_2-weighted image, magnetization is allowed to decay before measuring the MR signal by changing the echo time (TE). This image weighting is useful for detecting edema and inflammation, revealing white matter lesions, and assessing zonal anatomy in the prostate and uterus.

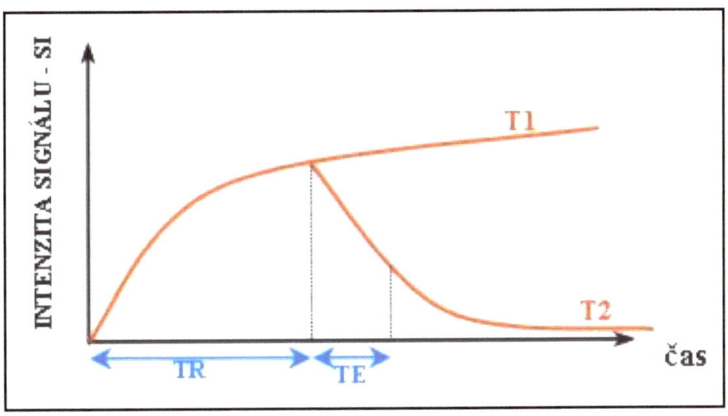

Fig: Effects of TR and TE on MR Signal

The information from MRI scans comes in the form of image contrasts based on differences in the rate of relaxation of nuclear spins following their perturbation by an oscillating magnetic field (in the form of radiofrequency pulses through the sample). The relaxation rates are a measure of the time it takes for a signal to decay back to an equilibrium state from either the longitudinal or transverse plane.

Magnetization builds up along the z-axis in the presence of a magnetic field, B_0, such that the magnetic dipoles in the sample will, on average, align with the z-axis summing to a total magnetization M_z. This magnetization along z is defined as the equilibrium magnetization; magnetization is defined as the sum of all magnetic dipoles in a sample. Following the equilibrium magnetization, a 90° radiofrequency

(RF) pulse flips the direction of the magnetization vector in the xy-plane, and is then switched off. The initial magnetic field B_0, however, is still applied. Thus, the spin magnetization vector will slowly return from the xy-plane back to the equilibrium state. The time it takes for the magnetization vector to return to its equilibrium value, M_z, is referred to as the longitudinal relaxation time, T_1. Subsequently, the rate at which this happens is simply the reciprocal of the relaxation time: $\frac{1}{T_1} = R_1$. Similarly, the time in which it takes for M_{xy} to return to zero is T_2, with the rate $\frac{1}{T_2} = R_2$. Magnetization as a function of time is defined by the Bloch equations.

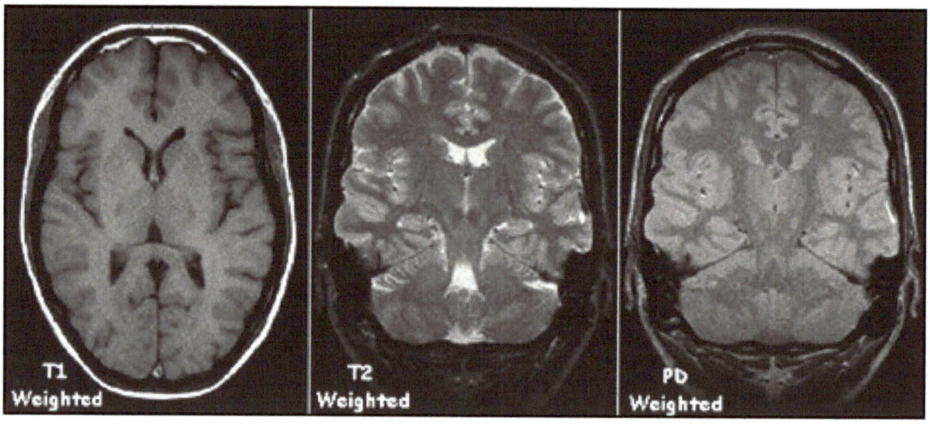

Fig: Examples of T1-Weighted, T2-Weighted and PD-Weighted MRI Scans

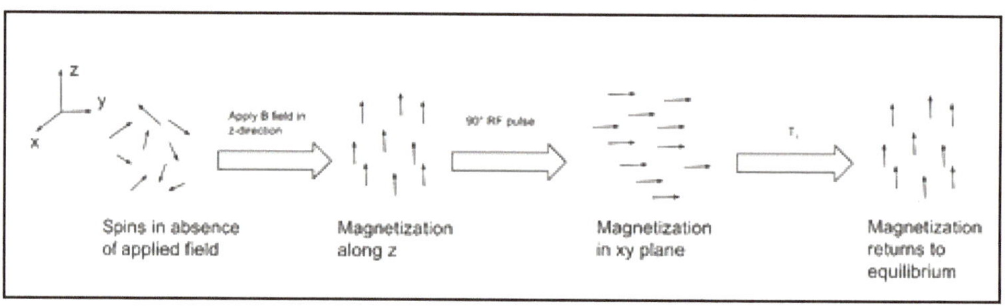

Fig: Diagram of Changing Magnetization and Spin Orientations throughout Spin-Lattice Relaxation Sxperiment

T_1 and T_2 values are dependent on the chemical environment of the sample; hence their utility in MRI. Soft tissue and muscle tissue relax at different rates, yielding the image contrast in a typical scan.

Usage by Organ or System

MRI has a wide range of applications in medical diagnosis and around 50,000 scanners are estimated to be in use worldwide. MRI affects diagnosis and treatment in many specialties although the effect on improved health outcomes is disputed in certain cases.

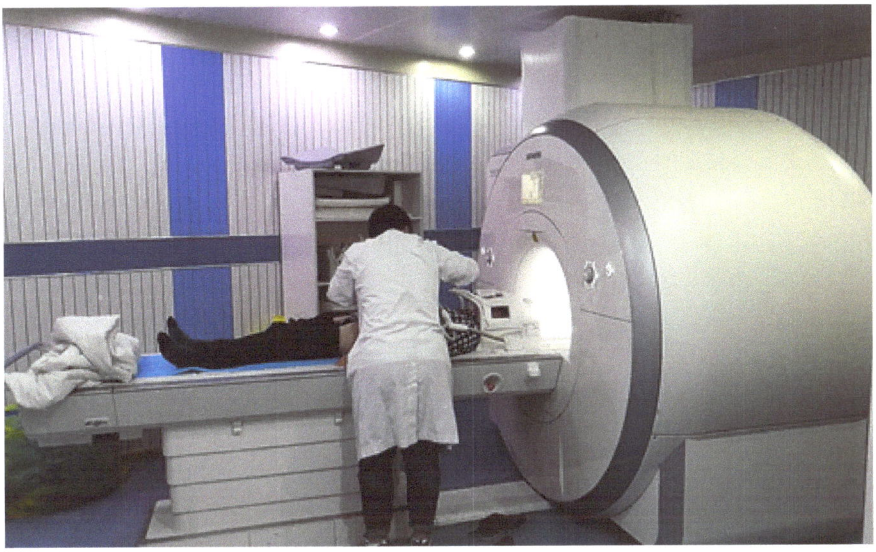

Fig: Patient Being Positioned for MR Study of the Head and Abdomen

MRI is the investigation of choice in the preoperative staging of rectal and prostate cancer and has a role in the diagnosis, staging, and follow-up of other tumors, as well as for determining areas of tissue for sampling in biobanking.

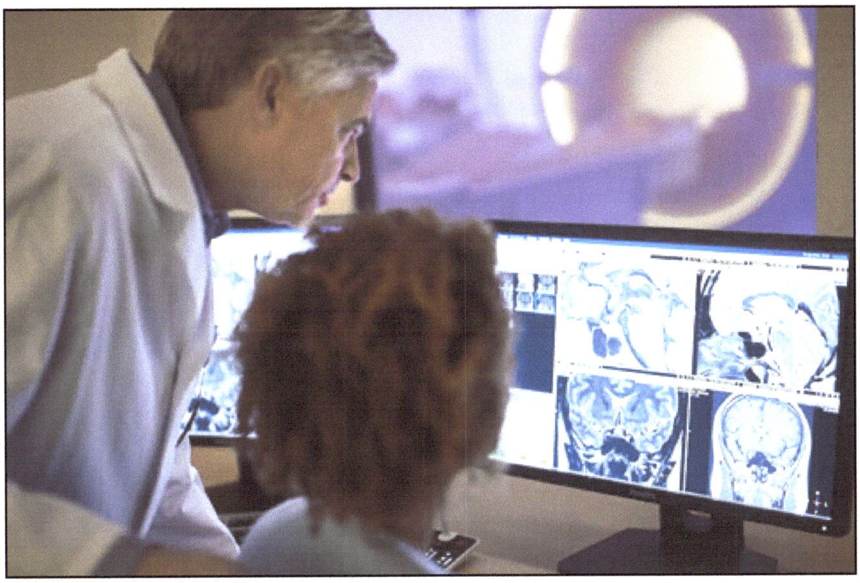

Fig: Radiologist Interpreting MRI Images of Head and Neck

Neuroimaging

MRI is the investigative tool of choice for neurological cancers over CT, as it offers better visualization of the posterior cranial fossa, containing the brainstem and the cerebellum. The contrast provided between grey and white matter makes

MRI the best choice for many conditions of the central nervous system, including demyelinating diseases, dementia, cerebrovascular disease, infectious diseases, Alzheimer's disease and epilepsy. Since many images are taken milliseconds apart, it shows how the brain responds to different stimuli, enabling researchers to study both the functional and structural brain abnormalities in psychological disorders. MRI also is used in guided stereotactic surgery and radiosurgery for treatment of intracranial tumors, arteriovenous malformations, and other surgically treatable conditions using a device known as the N-localizer. New tools that implement artificial intelligence in healthcare have demonstrated higher image quality and morphometric analysis in neuroimaging with the application of a denoising system.

The record for the highest spatial resolution of a whole intact brain (postmortem) is 100 microns, from Massachusetts General Hospital. The data was published in NATURE on 30 October 2019. Though MRI is used widely in research on mental disabilities, based on a 2024 systematic literature review and meta analysis commissioned by the Patient-Centered Outcomes Research Institute (PCORI), available research using MRI scans to diagnose ADHD showed great variability. The authors conclude that MRI cannot be reliably used to assist in making a clinical diagnosis of ADHD.

Cardiovascular

Cardiac MRI is complementary to other imaging techniques, such as echocardiography, cardiac CT, and nuclear medicine. It can be used to assess the structure and the function of the heart. Its applications include assessment of myocardial ischemia and viability, cardiomyopathies, myocarditis, iron overload, vascular diseases, and congenital heart disease.

Musculoskeletal

Applications in the musculoskeletal system include spinal imaging, assessment of joint disease, and soft tissue tumors. MRI techniques can also be used for diagnostic imaging of systemic muscle diseases including genetic muscle diseases. Swallowing movements of the throat and esophagus can cause motion artifacts over the imaged spine. Therefore, a saturation pulse applied over this region the throat and esophagus can help to avoid these artifacts. Motion artifacts arising due to pumping of the heart can be reduced by timing the MRI pulse according to heart cycles. Blood vessel flow artifacts can be reduced by applying saturation pulses above and below the region of interest.

Liver and Gastrointestinal

Hepatobiliary MR is used to detect and characterize lesions of the liver, pancreas, and bile ducts. Focal or diffuse disorders of the liver may be evaluated using diffusion-weighted, opposed-phase imaging and dynamic contrast enhancement sequences. Extracellular contrast agents are used widely in liver MRI, and newer hepatobiliary contrast agents also provide the opportunity to perform functional biliary imaging. Anatomical imaging of the bile ducts is achieved by using a heavily T2-weighted sequence in magnetic resonance cholangiopancreatography (MRCP). Functional imaging of the pancreas is performed following administration of secretin. MR enterography provides non-invasive assessment of inflammatory bowel disease and small bowel tumors. MR-colonography may play a role in the detection of large polyps in patients at increased risk of colorectal cancer.

Angiography

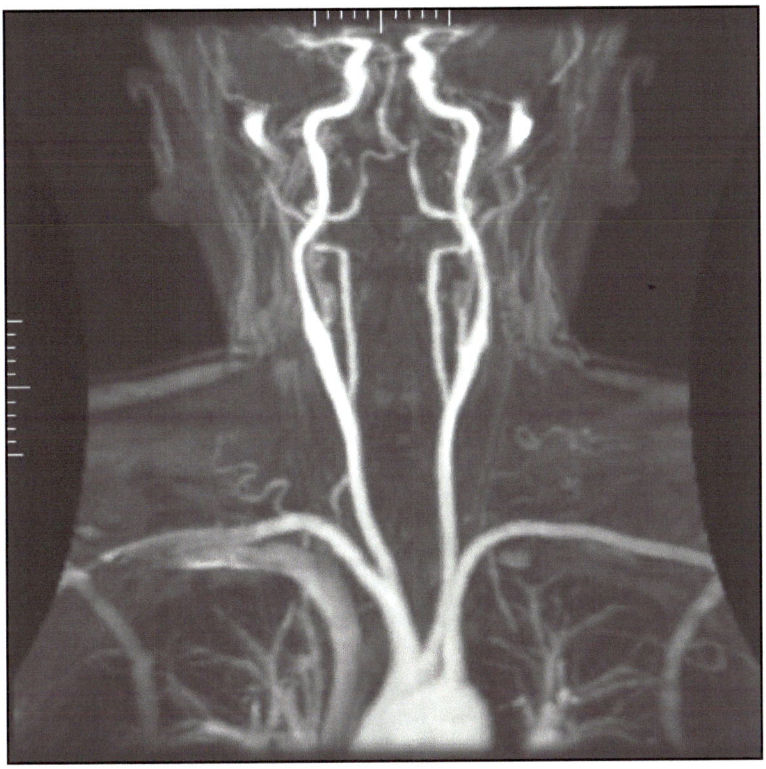

Fig: Magnetic Resonance Angiography

Magnetic resonance angiography (MRA) generates pictures of the arteries to evaluate them for stenosis (abnormal narrowing) or aneurysms (vessel wall dilatations, at risk of rupture). MRA is often used to evaluate the arteries of the neck and brain, the thoracic and abdominal aorta, the renal arteries, and the legs (called a "run-off"). A variety of techniques can be used to generate the pictures, such as administration of a paramagnetic contrast agent (gadolinium) or using a technique known as "flow-related enhancement" (e.g., 2D and 3D time-of-flight sequences), where most of the signal on an image is due to blood that recently moved into that plane.

Techniques involving phase accumulation (known as phase contrast angiography) can also be used to generate flow velocity maps easily and accurately. Magnetic resonance venography (MRV) is a similar procedure that is used to image veins. In this method, the tissue is now excited inferiorly, while the signal is gathered in the plane immediately superior to the excitation plane—thus imaging the venous blood that recently moved from the excited plane.

Contrast Agents

MRI for imaging anatomical structures or blood flow do not require contrast agents since the varying properties of the tissues or blood provide natural contrasts. However, for more specific types of imaging, exogenous contrast agents may be given intravenously, orally, or intra-articularly. Most contrast agents are either paramagnetic (e.g.: gadolinium, manganese, europium), and are used to shorten T1 in the tissue they accumulate in, or super-paramagnetic (SPIONs), and are used to shorten T2 and T2* in healthy tissue reducing its signal intensity (negative contrast agents). The most commonly used intravenous contrast agents are based on chelates of gadolinium, which is highly paramagnetic. In general, these agents have proved safer than the iodinated contrast agents used in X-ray radiography or CT. Anaphylactoid reactions are rare, occurring in approx. 0.03–0.1%. Of particular interest is the lower incidence of nephrotoxicity, compared with iodinated agents, when given at usual doses—this has made contrast-enhanced MRI scanning an option for patients with renal impairment, who would otherwise not be able to undergo contrast-enhanced CT.

Gadolinium-based contrast reagents are typically octadentate complexes of gadolinium(III). The complex is very stable (log K > 20) so that, in use, the concentration of the un-complexed Gd^{3+} ions should be below the toxicity limit. The 9th place in the metal ion's coordination sphere is occupied by a water molecule which exchanges rapidly with water molecules in the reagent molecule's immediate environment, affecting the magnetic resonance relaxation time.

In December 2017, the Food and Drug Administration (FDA) in the United States announced in a drug safety communication that new warnings were to be included on all gadolinium-based contrast agents (GBCAs). The FDA also called for increased patient education and requiring gadolinium contrast vendors to conduct additional animal and clinical studies to assess the safety of these agents. Although gadolinium agents have proved useful for patients with kidney impairment, in patients with severe kidney failure requiring dialysis there is a risk of a rare but serious illness, nephrogenic systemic fibrosis, which may be linked to the use of certain gadolinium-containing agents. The most frequently linked is gadodiamide, but other agents have been linked too. Although a causal link has not been definitively established, current guidelines in the United States are that dialysis patients should only receive gadolinium agents where essential and that dialysis should be performed as soon as possible after the scan to remove the agent from the body promptly.

In Europe, where more gadolinium-containing agents are available, a classification of agents according to potential risks has been released. In 2008, a new contrast agent named gadoxetate, brand name Eovist (US) or Primovist (EU), was approved for diagnostic use: This has the theoretical benefit of a dual excretion path.

Magnetic Resonance Spectroscopy

Magnetic resonance spectroscopy (MRS) is used to measure the levels of different metabolites in body tissues, which can be achieved through a variety of single voxel or imaging-based techniques. The MR signal produces a spectrum of resonances that corresponds to different molecular arrangements of the isotope being "excited". This signature is used to diagnose certain metabolic disorders, especially those affecting the brain, and to provide information on tumor metabolism. Magnetic resonance spectroscopic imaging (MRSI) combines both spectroscopic and imaging methods to produce spatially localized spectra from within the sample or patient. The spatial resolution is much lower (limited by the available SNR), but the spectra in each voxel contains information about many metabolites. Because the available signal is used to encode spatial and spectral information, MRSI requires high SNR achievable only at higher field strengths. The high procurement and maintenance costs of MRI with extremely high field strengths inhibit their popularity. However, recent compressed sensing-based software algorithms (*e.g.*, SAMV) have been proposed to achieve super-resolution without requiring such high field strengths.

Real-Time

Real-time magnetic resonance imaging (RT-MRI) refers to the continuous monitoring of moving objects in real time. Traditionally, real-time MRI was possible only with low image quality or low temporal resolution. An iterative reconstruction algorithm removed limitations. Radial FLASH MRI (real-time) yields a temporal resolution of 20 to 30 milliseconds for images with an in-plane resolution of 1.5 to 2.0 mm. Real-time MRI adds information about diseases of the joints and the heart. In many cases MRI examinations become easier and more comfortable for patients, especially for the patients who cannot calm their breathing or who have arrhythmia. Balanced steady-state free precession (bSSFP) imaging gives better image contrast between the blood pool and myocardium than FLASH MRI, at the cost of severe banding artifact when B0 inhomogeneity is strong.

Interventional MRI

The lack of harmful effects on the patient and the operator make MRI well-suited for interventional radiology, where the images produced by an MRI scanner guide minimally invasive procedures. Such procedures use no ferromagnetic instruments. A specialized growing subset of interventional MRI is intraoperative MRI, in which an MRI is used in surgery. Some specialized MRI systems allow imaging concurrent with the surgical procedure. More typically, the surgical procedure is temporarily interrupted so that MRI can assess the success of the procedure or guide subsequent surgical work.

Magnetic Resonance Guided Focused Ultrasound

In guided therapy, high-intensity focused ultrasound (HIFU) beams are focused on a tissue, that are controlled using MR thermal imaging. Due to the high energy at the focus, the temperature rises to above 65 °C (150 °F) which completely destroys the tissue. This technology can achieve precise ablation of diseased tissue. MR imaging provides a three-dimensional view of the target tissue, allowing for the precise focusing of ultrasound energy. The MR imaging provides quantitative, real-time, thermal images of the treated area. This allows the physician to ensure that the temperature generated during each cycle of ultrasound energy is sufficient to cause thermal ablation within the desired tissue and if not, to adapt the parameters to ensure effective treatment.

Multinuclear Imaging

Hydrogen has the most frequently imaged nucleus in MRI because it is present in biological tissues in great abundance, and because its high gyromagnetic ratio gives a strong signal. However, any nucleus with a net nuclear spin could potentially be imaged with MRI. Such nuclei include deuterium, helium-3, lithium-7, carbon-13, fluorine-19, oxygen-17, sodium-23, phosphorus-31 and xenon-129. ^{2}H, ^{23}Na and ^{31}P are naturally abundant in the body, so they can be imaged directly. Naturally abundant deuterium at the concentration of around 15mM can be imaged, but suffers from low gamma sensitivity and quadripolar Relaxation (NMR). Deuterium imaging however has a sparse chemical shift spectra making it possible to develop tailored multiband selective RF pulses for metabolite selective imaging. Thus, metabolic imaging, similar to what's done with Carbon-13 is possible with Deuterium metabolic imaging (DMI) for insights into vivo metabolic processes. As well, the short

T2 of deuterium allows it to be signal averaged rapidly, making up for some of its physical shortcomings. Gaseous isotopes such as ^{3}He or ^{129}Xe must be hyperpolarized and then inhaled as their nuclear density is too low to yield a useful signal under normal conditions. ^{17}O and ^{19}F can be administered in sufficient quantities in liquid form (e.g. ^{17}O-water) that hyperpolarization is not a necessity. Using helium or xenon has the advantage of reduced background noise, and therefore increased contrast for the image itself, because these elements are not normally present in biological tissues.

Moreover, the nucleus of any atom that has a net nuclear spin and that is bonded to a hydrogen atom could potentially be imaged via heteronuclear magnetization transfer MRI that would image the high-gyromagnetic-ratio hydrogen nucleus instead of the low-gyromagnetic-ratio nucleus that is bonded to the hydrogen atom. In principle, heteronuclear magnetization transfer MRI could be used to detect the presence or absence of specific chemical bonds.

Multinuclear imaging is primarily a research technique at present. However, potential applications include functional imaging and imaging of organs poorly seen on ^{1}H MRI (e.g., lungs and bones) or as alternative contrast agents. Inhaled hyperpolarized ^{3}He can be used to image the distribution of air spaces within the lungs. Injectable solutions containing ^{13}C or stabilized bubbles of hyperpolarized ^{129}Xe have been studied as contrast agents for angiography and perfusion imaging. ^{31}P can potentially provide information on bone density and structure, as well as functional imaging of the brain. Multinuclear imaging holds the potential to chart the distribution of lithium in the human brain, this element finding use as an important drug for those with conditions such as bipolar disorder.

Molecular Imaging by MRI

MRI has the advantages of having very high spatial resolution and is very adept at morphological imaging and functional imaging. MRI does have several disadvantages though. First, MRI has a sensitivity of around 10^{-3} mol/L to 10^{-5} mol/L, which, compared to other types of imaging, can be very limiting. This problem stems from the fact that the population difference between the nuclear spin states is very small at room temperature. For example, at 1.5 teslas, a typical field strength for clinical MRI, the difference between high and low energy states is approximately 9 molecules per 2 million. Improvements to increase MR sensitivity include increasing magnetic field strength and hyperpolarization via optical pumping or dynamic nuclear polarization. There are also a variety of signal amplification schemes based on chemical exchange that increase sensitivity.

To achieve molecular imaging of disease biomarkers using MRI, targeted MRI contrast agents with high specificity and high relaxivity (sensitivity) are required. To date, many studies have been devoted to developing targeted-MRI contrast agents to achieve molecular imaging by MRI. Commonly, peptides, antibodies, or small ligands, and small protein domains, such as HER-2 affibodies, have been applied to achieve targeting.

To enhance the sensitivity of the contrast agents, these targeting moieties are usually linked to high payload MRI contrast agents or MRI contrast agents with high relaxivities. A new class of gene targeting MR contrast agents has been introduced to show gene action of unique mRNA and gene transcription factor proteins. These new contrast agents can trace cells with unique mRNA, microRNA and virus; tissue response to inflammation in living brains. The MR reports change in gene expression with positive correlation to TaqMan analysis, optical and electron microscopy.

Parallel MRI

It takes time to gather MRI data using sequential applications of magnetic field gradients. Even for the most streamlined of MRI sequences, there are physical and physiologic limits to the rate of gradient switching. Parallel MRI circumvents these limits by gathering some portion of the data simultaneously, rather than in a traditional sequential fashion. This is accomplished using arrays of radiofrequency (RF) detector coils, each with a different 'view' of the body. A reduced set of gradient steps is applied, and the remaining spatial information is filled in by combining signals from various coils, based on their known spatial sensitivity patterns. The resulting acceleration is limited by the number of coils and by the signal to noise ratio (which decreases with increasing acceleration), but two- to four-fold accelerations may commonly be achieved with suitable coil array configurations, and substantially higher accelerations have been demonstrated with specialized coil arrays. Parallel MRI may be used with most MRI sequences.

After a number of early suggestions for using arrays of detectors to accelerate imaging went largely unremarked in the MRI field, parallel imaging saw widespread development and application following the introduction of the Simultaneous Acquisition of Spatial Harmonics (SMASH) technique in 1996–7. The Sensitivity Encoding (SENSE) and Generalized Autocalibrating Partially Parallel Acquisitions (GRAPPA) techniques are the parallel imaging methods in most common use today. The advent of parallel MRI resulted in extensive research and development in image reconstruction and RF coil design, as well as in a rapid expansion of the number of receiver channels available on commercial MR systems. Parallel MRI is now used routinely for MRI examinations in a wide range of body areas and clinical or research applications.

Quantitative MRI

Most MRI focuses on qualitative interpretation of MR data by acquiring spatial maps of relative variations in signal strength which are "weighted" by certain parameters. Quantitative methods instead attempt to determine spatial maps of accurate tissue relaxometry parameter values or magnetic field, or to measure the size of certain spatial features.

Examples of quantitative MRI methods are:

- T1-mapping (notably used in cardiac magnetic resonance imaging)
- T2-mapping

- Quantitative susceptibility mapping (QSM)
- Quantitative fluid flow MRI (i.e. some cerebrospinal fluid flow MRI)
- Magnetic resonance elastography (MRE)

Quantitative MRI aims to increase the reproducibility of MR images and interpretations, but has historically require longer scan times. Quantitative MRI (or qMRI) sometimes more specifically refers to multi-parametric quantitative MRI, the mapping of multiple tissue relaxometry parameters in a single imaging session. Efforts to make multi-parametric quantitative MRI faster have produced sequences which map multiple parameters simultaneously, either by building separate encoding methods for each parameter into the sequence, or by fitting MR signal evolution to a multi-parameter model.

Hyperpolarized Gas MRI

Traditional MRI generates poor images of lung tissue because there are fewer water molecules with protons that can be excited by the magnetic field. Using hyperpolarized gas an MRI scan can identify ventilation defects in the lungs. Before the scan, a patient is asked to inhale hyperpolarized xenon mixed with a buffer gas of helium or nitrogen. The resulting lung images are much higher quality than with traditional MRI.

Safety

MRI is, in general, a safe technique, although injuries may occur as a result of failed safety procedures or human error. Contraindications to MRI include most cochlear implants and cardiac pacemakers, shrapnel, and metallic foreign bodies in the eyes. Magnetic resonance imaging in pregnancy appears to be safe, at least during the second and third trimesters if done without contrast agents. Since MRI does not use any ionizing radiation, its use is generally favored in preference to CT when either modality could yield the same information. Some patients experience claustrophobia and may require sedation or shorter MRI protocols. Amplitude and rapid switching of gradient coils during image acquisition may cause peripheral nerve stimulation.

MRI uses powerful magnets and can therefore cause magnetic materials to move at great speeds, posing a projectile risk, and may cause fatal accidents. However, as millions of MRIs are performed globally each year, fatalities are extremely rare. MRI machines can produce loud noise, up to 120 dB(A). This can cause hearing loss, tinnitus and hyperacusis, so appropriate hearing protection is essential for anyone inside the MRI scanner room during the examination.

- Overuse: Medical societies issue guidelines for when physicians should use MRI on patients and recommend against overuse. MRI can detect health problems or confirm a diagnosis, but medical societies often recommend that MRI not be the first procedure for creating a plan to diagnose or manage a patient's complaint. A common case is to use MRI to seek a cause of low back pain; the American College of Physicians, for example, recommends against imaging (including MRI) as unlikely to result in a positive outcome for the patient.

Artifacts

An MRI artifact is a visual artifact, that is, an anomaly during visual representation. Many different artifacts can occur during magnetic resonance imaging (MRI), some affecting the diagnostic quality, while others may be confused with pathology. Artifacts can be classified as patient-related, signal processing-dependent and hardware (machine)-related.

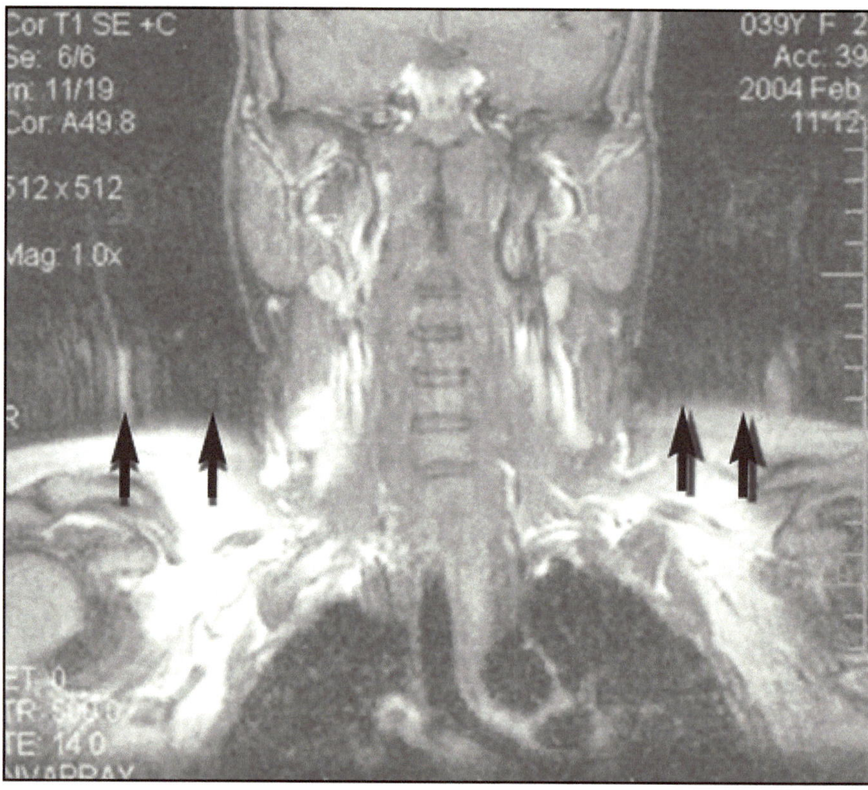

Fig: Motion Artifact (T1 Coronal Study of Cervical Vertebrae)

Non-Medical Use

MRI is used industrially mainly for routine analysis of chemicals. The nuclear magnetic resonance technique is also used, for example, to measure the ratio between water and fat in foods, monitoring of flow of corrosive fluids in pipes, or to study molecular structures such as catalysts. Being non-invasive and non-damaging, MRI can be used to study the anatomy of plants, their water transportation processes and water balance. It is also applied to veterinary radiology for diagnostic purposes. Outside this, its use in zoology is limited due to the high cost; but it can be used on many species. In palaeontology it is used to examine the structure of fossils. Forensic imaging provides graphic documentation of an autopsy, which manual autopsy does not. CT scanning provides quick whole-body imaging of skeletal and parenchymal alterations, whereas MR imaging gives better representation of soft tissue pathology. All that being said, MRI is more expensive, and more time-consuming to utilize. Moreover, the quality of MR imaging deteriorates below 10 °C.

Quantum Sensors for Enhancing MRI Sensitivity

One of the main challenges in MRI is detecting the weak magnetic signals produced by hydrogen nuclei in the body. These signals are often obscured by noise, leading to a reduced signal-to-noise ratio (SNR) and limiting the quality of the resulting images. Quantum sensors, specifically quantum magnetometers, have the potential to enhance the sensitivity of MRI systems by detecting magnetic fields with much higher precision than classical sensors.

Quantum magnetometers, such as those based on nitrogen-vacancy (NV) centers in diamonds, exploit the quantum properties of these defects to detect extremely weak magnetic fields. NV centers are atomic-scale defects in the crystal lattice of diamonds that can be manipulated and measured using quantum mechanical properties. By applying this technology to MRI, quantum sensors can provide more precise measurements of the magnetic fields generated by hydrogen nuclei in the body, leading to higher SNR and better image quality.

Another promising technology for enhancing MRI sensitivity is the use of quantum-enhanced sensors based on superconducting qubits, which exploit the principles of quantum superposition and entanglement. Superconducting qubits can be used to measure the weak magnetic fields in the body with high sensitivity, offering a potential solution to the low SNR problem in MRI. These quantum sensors can also be integrated with existing MRI systems, providing a seamless upgrade to improve image quality and reduce scan times.

Improved Spatial Resolution with Quantum Sensors

In addition to enhancing sensitivity, quantum sensors also have the potential to improve the spatial resolution of MRI scans. Traditional MRI systems rely on magnetic gradients to encode spatial information into the signals, but these gradients are limited in their ability to achieve high resolution without increasing scan time. Quantum sensors can provide a more precise measurement of magnetic fields, allowing for finer spatial resolution without the need for longer scan times.

Quantum-enhanced spatial resolution could enable the imaging of smaller structures in the body, such as individual cells or small tumors, with greater accuracy. This would be particularly beneficial in the early detection of diseases such as cancer, where the ability to detect small abnormalities could significantly improve patient outcomes. Additionally, the increased resolution could allow for the visualization of complex structures in the brain or cardiovascular system, aiding in the diagnosis of neurological or heart-related conditions.

Faster MRI Scans with Quantum Sensors

One of the significant benefits of quantum sensors in MRI is the potential to reduce scan times while maintaining or even improving image quality. Traditional MRI scans can take anywhere from 20 minutes to over an hour, depending on the area being imaged and the resolution required. Longer scan times are often necessary to improve image quality and sensitivity, but this can be uncomfortable for patients and may not be ideal for certain medical conditions. Quantum sensors, by improving the signal-to-noise ratio and spatial resolution, can potentially reduce the need for long scan times. For example, the use of quantum-enhanced magnetometers, such as NV centers in diamonds, can improve the sensitivity of the MRI system without requiring longer scans. By achieving better image quality more quickly, quantum sensors could make MRI a more efficient and patient-friendly procedure, allowing for faster diagnoses and improved clinical outcomes.

Quantum Sensors in Functional MRI (fMRI)

Functional MRI (fMRI) is a specialized form of MRI that measures brain activity by detecting changes in blood oxygen levels, which are linked to neural activity. fMRI has been instrumental in understanding brain function, but it is limited by the sensitivity of the MRI system and the need for higher spatial resolution to map brain activity accurately. Quantum sensors have the potential to enhance fMRI by providing higher sensitivity to detect smaller fluctuations in magnetic fields associated with brain activity.

By utilizing quantum magnetometers, fMRI could achieve more precise measurements of neural activity and provide higher-resolution images of brain function. This could lead to breakthroughs in understanding neurological diseases such as Alzheimer's, Parkinson's, and epilepsy. Quantum sensors could also improve the spatial resolution of fMRI scans, enabling the mapping of brain regions at a much finer level, potentially leading to more effective treatments and therapies for various neurological conditions.

Challenges and Future Prospects

While the potential benefits of quantum sensors for MRI enhancement are clear, there are still several challenges that need to be addressed before these technologies can be widely adopted in clinical settings. One of the primary challenges is the integration of quantum sensors into existing MRI systems. Quantum sensors, such as NV centers and superconducting qubits, are still in the experimental stages, and their integration into large-scale, high-throughput medical imaging systems will require significant advancements in both hardware and software.

Additionally, the need for ultra-cold environments and complex quantum control mechanisms could pose logistical challenges for integrating quantum sensors into clinical MRI systems. Developing more user-friendly and cost-effective quantum sensors will be crucial to making this technology accessible to hospitals and medical centers worldwide.

Despite these challenges, the future of quantum-enhanced MRI is promising. Ongoing research and development in quantum technologies are expected to lead to significant breakthroughs in sensor sensitivity, resolution, and speed. As quantum sensors become more refined and accessible, their integration into MRI systems could revolutionize medical imaging, enabling faster, more accurate diagnoses and improving patient outcomes across a wide range of medical conditions.

5.3.2 Quantum Sensors for Biomolecular Sensing

Quantum sensors represent a new frontier in sensing technologies, offering unprecedented sensitivity and precision by leveraging the principles of quantum mechanics. Their application in biomolecular sensing is especially promising, given the need for highly sensitive, label-free, and non-invasive detection methods in molecular biology, biomedical diagnostics, and drug development. Traditional methods of biomolecular sensing, such as fluorescence tagging, surface plasmon resonance, or electrochemical techniques, often require complex sample preparation or are limited by detection thresholds. In contrast, quantum sensors can detect minute changes in physical quantities like magnetic fields, electric fields, and temperature, allowing them to monitor biological phenomena at the molecular or even atomic scale.

This capability opens up new possibilities for early disease detection, real-time monitoring of cellular processes, and the development of advanced biosensors for research and clinical use. Quantum sensors are poised to revolutionize the field of biomolecular sensing by offering unmatched sensitivity, spatial resolution, and versatility. Their ability to detect single molecules, monitor real-time biological interactions, and operate under ambient conditions makes them ideal for applications ranging from fundamental research to clinical diagnostics. By exploiting quantum mechanical phenomena such as spin coherence and entanglement, these sensors provide powerful new tools for exploring biological systems at the nanoscale. As the technology matures and overcomes existing challenges, quantum sensors will likely become a cornerstone of next-generation biosensing platforms, transforming how we detect, understand, and treat diseases.

Principles of Quantum Sensing

Quantum sensors operate based on the unique properties of quantum systems, such as superposition, entanglement, and coherence. These properties enable quantum sensors to respond to extremely subtle environmental changes. One widely studied quantum system for sensing is the nitrogen-vacancy (NV) center in diamond, which can detect magnetic fields, electric fields, and temperature changes with high spatial resolution. In the context of biomolecular sensing, these quantum systems can interact with the weak magnetic fields generated by biological molecules or detect changes in their physical surroundings caused by molecular binding events. The high sensitivity and spatial resolution of quantum sensors allow them to perform measurements that surpass the standard quantum limit, making them ideally suited for tasks that require the detection of very small numbers of molecules or subtle biochemical changes.

Nitrogen-Vacancy Centers and Their Role in Biosensing

Nitrogen-vacancy centers in diamond have emerged as one of the most promising platforms for quantum sensing in biological applications. These point defects in the diamond lattice consist of a nitrogen atom adjacent to a vacancy in the carbon lattice. NV centers exhibit unique spin properties that can be optically initialized and read out at room temperature. The quantum spin states of NV centers are highly sensitive to magnetic and electric fields, temperature, and strain, enabling them to function as versatile sensors for a wide range of environmental parameters. In biomolecular sensing, NV centers can be brought into close proximity with biological samples, such as proteins, nucleic acids, or lipid membranes. When molecules interact or undergo conformational changes, they often generate minute local magnetic or electric fields, which can be detected by the NV centers. This allows for label-free detection of binding events or conformational dynamics, which is particularly valuable in applications where traditional fluorescent tags may interfere with biological processes. NV-based sensors have also been used to detect paramagnetic ions and free radicals in biological environments, contributing to the study of oxidative stress and enzymatic activity at the nanoscale.

Quantum Sensors for Single-Molecule Detection

One of the most compelling applications of quantum sensors in biomolecular research is their ability to detect and analyze single molecules. Single-molecule detection provides deep insights into the stochastic behaviors and heterogeneous dynamics of biomolecules, which are often averaged out in bulk measurements. Quantum sensors, particularly NV centers, can detect the magnetic noise produced by single electron or nuclear spins, making them powerful tools for studying molecular interactions at the single-molecule level. In experimental setups, NV centers embedded in nanodiamonds or near the diamond surface can be used to probe molecules tethered on substrates or suspended in solution. The quantum sensor measures changes in the local magnetic field as a result of the presence or movement of a biomolecule, providing real-time data on binding events, enzymatic reactions, or conformational changes. Such high-resolution sensing is crucial for understanding fundamental biological mechanisms, such as protein folding, ligand-receptor interactions, and DNA hybridization, and it paves the way for early diagnostics based on single-molecule biomarkers.

Quantum Magnetometry in Biological Environments

Quantum magnetometry is a technique that measures extremely small magnetic fields using quantum systems. In the context of biomolecular sensing, it allows the detection of magnetic fields associated with biological molecules or processes, such as electron spin transitions in metal cofactors, or the magnetic fields generated by neural activity or cardiac signals. This application is particularly relevant in detecting redox reactions and catalytic activity of metalloenzymes, which are essential in metabolic pathways. Biological environments are often noisy and complex, with thermal and electromagnetic fluctuations that can interfere with sensitive measurements.

However, quantum sensors such as NV centers in diamonds maintain coherence even in noisy surroundings, enabling accurate and reliable measurements. Their robustness and compatibility with ambient conditions make them suitable for use in physiological and in vivo settings, where traditional sensors may struggle to perform. These capabilities can lead to the development of portable and highly sensitive diagnostic tools capable of detecting disease biomarkers in real-time and at ultra-low concentrations.

Applications in Drug Discovery and Screening

Quantum sensors also show significant promise in pharmaceutical research, particularly in drug discovery and high-throughput screening. Conventional methods for drug screening often rely on bulk assays, which may not accurately reflect the interactions occurring at the molecular level. Quantum sensors, with their ability to detect and monitor individual molecular interactions, offer a more nuanced understanding of how drugs bind to their targets, how long they remain bound, and how they affect molecular dynamics. For instance, by observing magnetic noise and spin fluctuations associated with biomolecular binding, quantum sensors can provide kinetic and thermodynamic information about drug-target interactions. This enables more precise characterization of drug efficacy, binding affinity, and specificity. Furthermore, the use of quantum sensors in high-throughput formats, such as diamond arrays with multiple NV centers, could significantly speed up the drug screening process while reducing costs and improving data quality.

Integration with Microfluidics and Lab-on-a-Chip Technologies

To harness the full potential of quantum sensors in biomolecular sensing, researchers are integrating these sensors with microfluidic platforms and lab-on-a-chip devices. Such integration allows for the controlled manipulation and delivery of biological samples in a miniaturized environment, enhancing the sensitivity and throughput of quantum sensing experiments. Microfluidic channels can guide molecules close to the sensing region, where quantum measurements can be performed in real time. This combination is particularly advantageous for point-of-care diagnostics, where compact, low-power, and highly sensitive devices are required. Quantum sensors embedded in microfluidic chips could be used to detect biomarkers in blood, saliva, or urine samples without the need for extensive sample preparation or laboratory infrastructure. As such technologies continue to evolve, they could transform healthcare by enabling early disease detection and personalized medicine at the point of need.

Challenges and Future Outlook

Despite the exciting potential of quantum sensors in biomolecular sensing, several challenges remain. One major hurdle is the fabrication and integration of quantum sensors with biological systems in a reproducible and scalable manner. Ensuring biocompatibility, stability, and sensitivity of quantum sensors under physiological conditions is critical for their successful application. Additionally, interpreting quantum sensor signals in complex biological environments requires sophisticated algorithms and data analysis techniques, which are still under active development. Another challenge is miniaturization and cost-effectiveness. While laboratory demonstrations of quantum sensing are impressive, transitioning these technologies into affordable and user-friendly devices for widespread clinical or research use requires further engineering advancements. Progress in material science, quantum control techniques, and device fabrication will be essential in overcoming these barriers. Looking ahead, the future of quantum sensors in biomolecular sensing appears bright. As interdisciplinary collaboration between physicists, chemists, biologists, and engineers continues to grow, it is expected that quantum sensors will become integral tools in biomedical research, diagnostics, and therapy monitoring. Their ability to probe biological systems with unparalleled sensitivity and precision will unlock new insights into life at the molecular level and contribute to the advancement of personalized and precision medicine.

5.4 ENVIRONMENTAL SENSING WITH QUANTUM TECHNOLOGIES

5.4.1 Remote Sensing of Magnetic Fields and Temperature

The Basic of Remote Sensing

Remote sensing is the acquisition of information about an object or phenomenon without making physical contact with the object, in contrast to in situ or on-site observation. The term is applied especially to acquiring information about Earth and other planets. Remote sensing is used in numerous fields, including geophysics, geography, land surveying and most Earth science disciplines (e.g. exploration geophysics, hydrology, ecology, meteorology, oceanography, glaciology, geology). It also has military, intelligence, commercial, economic, planning, and humanitarian applications, among others.

In current usage, the term *remote sensing* generally refers to the use of satellite- or aircraft-based sensor technologies to detect and classify objects on Earth. It includes the surface and the atmosphere and oceans, based on propagated signals (e.g. electromagnetic radiation). It may be split into "active" remote sensing (when a signal is emitted by a satellite or aircraft to the object and its reflection is detected by the sensor) and "passive" remote sensing (when the reflection of sunlight is detected by the sensor).

Remote sensing can be divided into two types of methods: *Passive remote sensing* and *Active remote sensing.* Passive sensors gather radiation that is emitted or reflected by the object or surrounding areas. Reflected sunlight is the most common source of radiation measured by passive sensors. Examples of passive remote sensors include film photography, infrared, charge-coupled devices, and radiometers. Active collection, on the other hand, emits energy in order to scan objects and areas whereupon a sensor then detects and measures the radiation that is reflected or backscattered from the target. RADAR and LiDAR are examples of active remote sensing where the time delay between emission and return is measured, establishing the location, speed and direction of an object.

Remote sensing makes it possible to collect data of dangerous or inaccessible areas. Remote sensing applications include monitoring deforestation in areas such as the Amazon Basin, glacial features in Arctic and Antarctic regions, and depth sounding of coastal and ocean depths. Military collection during the Cold War made use of stand-off collection of data about dangerous border areas. Remote sensing also replaces costly and slow data collection on the ground, ensuring in the process that areas or objects are not disturbed.

Orbital platforms collect and transmit data from different parts of the electromagnetic spectrum, which in conjunction with larger scale aerial or ground-based sensing and analysis, provides researchers with enough information to monitor trends such as El Niño and other natural long and short term phenomena. Other uses include different areas of the earth sciences such as natural resource management, agricultural fields such as land usage and conservation, greenhouse gas monitoring, oil spill detection and monitoring, and national security and overhead, ground-based and stand-off collection on border areas.

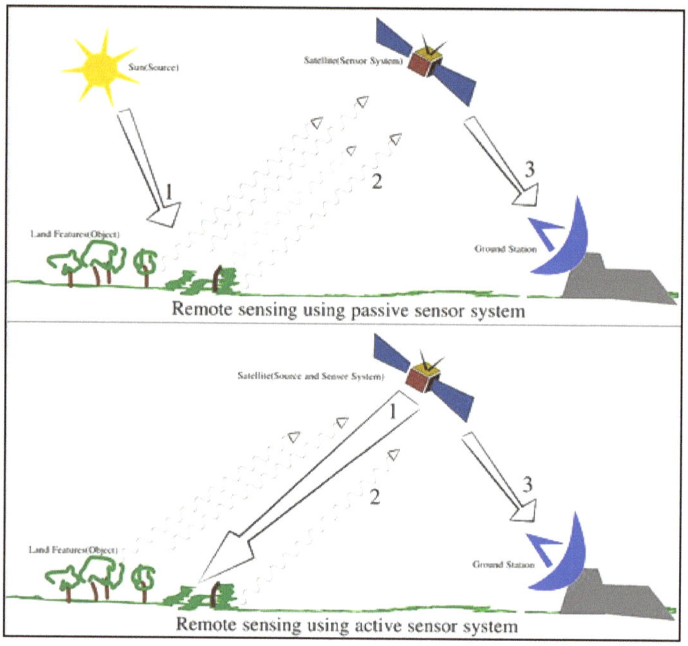

Fig: Illustration of Remote Sensing

Applications of Remote Sensing

- Conventional radar is mostly associated with aerial traffic control, early warning, and certain large-scale meteorological data. Doppler radar is used by local law enforcements' monitoring of speed limits and in enhanced meteorological collection such as wind speed and direction within weather systems in addition to precipitation location and intensity. Other types of active collection includes plasmas in the ionosphere. Interferometric synthetic aperture radar is used to produce precise digital elevation models of large scale terrain.
- Laser and radar altimeters on satellites have provided a wide range of data. By measuring the bulges of water caused by gravity, they map features on the seafloor to a resolution of a mile or so. By measuring the height and wavelength of ocean waves, the altimeters measure wind speeds and direction, and surface ocean currents and directions.
- Ultrasound (acoustic) and radar tide gauges measure sea level, tides and wave direction in coastal and offshore tide gauges.

- Light detection and ranging (LIDAR) is used for weapon ranging, laser illuminated homing of projectiles, and to detect and measure the concentration of various chemicals in the atmosphere while airborne LIDAR can be used to measure the heights of objects and features on the ground more accurately than radar technology. LIDAR can be used to detect ground surface changes. Vegetation remote sensing is a principal application of LIDAR.
- Radiometers and photometers are the most common instrument in use, collecting reflected and emitted radiation in a wide range of frequencies. The most common are visible and infrared sensors, followed by microwave, gamma-ray, and rarely, ultraviolet. They may also be used to detect the emission spectra of various chemicals, providing data on chemical concentrations in the atmosphere.

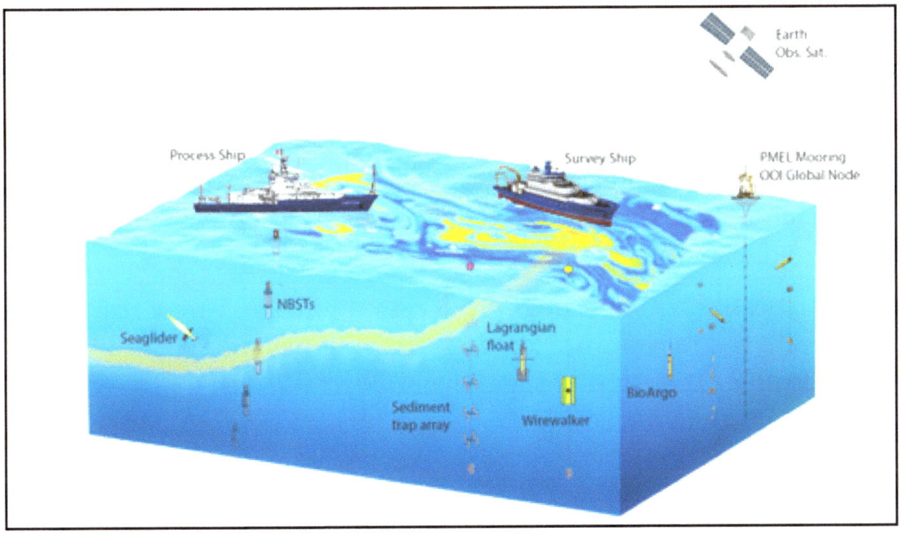

Fig: Examples of Remote Sensing Equipment Deployed by or Interfaced with Oceanographic Research Vessels

- Radiometers are also used at night, because artificial light emissions are a key signature of human activity. Applications include remote sensing of population, GDP, and damage to infrastructure from war or disasters.
- Radiometers and radar onboard of satellites can be used to monitor volcanic eruptions
- Spectropolarimetric Imaging has been reported to be useful for target tracking purposes by researchers at the U.S. Army Research Laboratory. They determined that manmade items possess polarimetric signatures that are not found in natural objects. These conclusions were drawn from the imaging of military trucks, like the Humvee, and trailers with their

acousto-optic tunable filter dual hyperspectral and spectropolarimetric VNIR Spectropolarimetric Imager.

- Stereographic pairs of aerial photographs have often been used to make topographic maps by imagery and terrain analysts in trafficability and highway departments for potential routes, in addition to modelling terrestrial habitat features.
- Simultaneous multi-spectral platforms such as Landsat have been in use since the 1970s. These thematic mappers take images in multiple wavelengths of electromagnetic radiation (multi-spectral) and are usually found on Earth observation satellites, including (for example) the Landsat program or the IKONOS satellite. Maps of land cover and land use from thematic mapping can be used to prospect for minerals, detect or monitor land usage, detect invasive vegetation, deforestation, and examine the health of indigenous plants and crops (satellite crop monitoring), including entire farming regions or forests. Prominent scientists using remote sensing for this purpose include Janet Franklin and Ruth DeFries. Landsat images are used by regulatory agencies such as KYDOW to indicate water quality parameters including Secchi depth, chlorophyll density, and total phosphorus content. Weather satellites are used in meteorology and climatology.
- Hyperspectral imaging produces an image where each pixel has full spectral information with imaging narrow spectral bands over a contiguous spectral range. Hyperspectral imagers are used in various applications including mineralogy, biology, defence, and environmental measurements.
- Within the scope of the combat against desertification, remote sensing allows researchers to follow up and monitor risk areas in the long term, to determine desertification factors, to support decision-makers in defining relevant measures of environmental management, and to assess their impacts.
- Remotely sensed multi- and hyperspectral images can be used for assessing biodiversity at different scales. Since the spectral properties of different plants species are unique, it is possible to get information about properties that relates to biodiversity such as habitat heterogeneity, spectral diversity and plant functional trait.
- Remote sensing has been used to detect rare plants to aid in conservation efforts. Prediction, detection, and the ability to record biophysical conditions were possible from medium to very high resolutions.

- Agricultural and environmental statistics, usually combining classified satellite images with ground truth data collected on a sample selected on an area sampling frame

Geodetic

- Geodetic remote sensing can be gravimetric or geometric. Overhead gravity data collection was first used in aerial submarine detection. This data revealed minute perturbations in the Earth's gravitational field that may be used to determine changes in the mass distribution of the Earth, which in turn may be used for geophysical studies, as in GRACE. Geometric remote sensing includes position and deformation imaging using InSAR, LIDAR, etc.

Acoustic and Near-Acoustic

- **Sonar:** *Passive sonar*, listening for the sound made by another object (a vessel, a whale etc.); *active sonar*, emitting pulses of sounds and listening for echoes, used for detecting, ranging and measurements of underwater objects and terrain.
- Seismograms taken at different locations can locate and measure earthquakes (after they occur) by comparing the relative intensity and precise timings.
- **Ultrasound:** Ultrasound sensors, that emit high-frequency pulses and listening for echoes, used for detecting water waves and water level, as in tide gauges or for towing tanks.

To coordinate a series of large-scale observations, most sensing systems depend on the following: platform location and the orientation of the sensor. High-end instruments now often use positional information from satellite navigation systems. The rotation and orientation are often provided within a degree or two with electronic compasses. Compasses can measure not just azimuth (i. e. degrees to magnetic north), but also altitude, since the magnetic field curves into the Earth at different angles at different latitudes. More exact orientations require gyroscopic-aided orientation, periodically realigned by different methods including navigation from stars or known benchmarks.

Data Characteristics

The quality of remote sensing data consists of its spatial, spectral, radiometric and temporal resolutions.

- **Spatial Resolution:** The size of a pixel that is recorded in a raster image – typically pixels may correspond to square areas ranging in side length from 1 to 1,000 metres (3.3 to 3,280.8 ft).

- **Spectral Resolution:** The wavelength of the different frequency bands recorded – usually, this is related to the number of frequency bands recorded by the platform. Current Landsat collection is that of seven bands, including several in the infrared spectrum, ranging from a spectral resolution of 0.7 to 2.1 μm. The Hyperion sensor on Earth Observing-1 resolves 220 bands from 0.4 to 2.5 μm, with a spectral resolution of 0.10 to 0.11 μm per band.
- **Radiometric Resolution:** The number of different intensities of radiation the sensor is able to distinguish. Typically, this ranges from 8 to 14 bits, corresponding to 256 levels of the gray scale and up to 16,384 intensities or "shades" of colour, in each band. It also depends on the instrument noise.
- **Temporal Resolution:** The frequency of flyovers by the satellite or plane, and is only relevant in time-series studies or those requiring an averaged or mosaic image as in deforesting monitoring. This was first used by the intelligence community where repeated coverage revealed changes in infrastructure, the deployment of units or the modification/introduction of equipment. Cloud cover over a given area or object makes it necessary to repeat the collection of said location.

Data Processing

In order to create sensor-based maps, most remote sensing systems expect to extrapolate sensor data in relation to a reference point including distances between known points on the ground. This depends on the type of sensor used. For example, in conventional photographs, distances are accurate in the center of the image, with the distortion of measurements increasing the farther you get from the center.

Another factor is that of the platen against which the film is pressed can cause severe errors when photographs are used to measure ground distances. The step in which this problem is resolved is called georeferencing and involves computer-aided matching of points in the image (typically 30 or more points per image) which is extrapolated with the use of an established benchmark, "warping" the image to produce accurate spatial data. As of the early 1990s, most satellite images are sold fully georeferenced.

In addition, images may need to be radiometrically and atmospherically corrected.

- **Radiometric Correction:** Allows avoidance of radiometric errors and distortions. The illumination of objects on the Earth's surface is uneven because of different properties of the relief. This factor is taken into account in the method of radiometric distortion correction. Radiometric correction gives a scale to the pixel values, e. g. the monochromatic scale of 0 to 255 will be converted to actual radiance values.

- **Topographic Correction (also Called Terrain Correction):** In rugged mountains, as a result of terrain, the effective illumination of pixels varies considerably. In a remote sensing image, the pixel on the shady slope receives weak illumination and has a low radiance value, in contrast, the pixel on the sunny slope receives strong illumination and has a high radiance value. For the same object, the pixel radiance value on the shady slope will be different from that on the sunny slope. Additionally, different objects may have similar radiance values. These ambiguities seriously affected remote sensing image information extraction accuracy in mountainous areas. It became the main obstacle to the further application of remote sensing images. The purpose of topographic correction is to eliminate this effect, recovering the true reflectivity or radiance of objects in horizontal conditions. It is the premise of quantitative remote sensing application.
- **Atmospheric Correction:** Elimination of atmospheric haze by rescaling each frequency band so that its minimum value (usually realised in water bodies) corresponds to a pixel value of 0. The digitizing of data also makes it possible to manipulate the data by changing gray-scale values.

Interpretation is the critical process of making sense of the data. The first application was that of aerial photographic collection which used the following process; spatial measurement through the use of a light table in both conventional single or stereographic coverage, added skills such as the use of photogrammetry, the use of photomosaics, repeat coverage, Making use of objects' known dimensions in order to detect modifications. Image Analysis is the recently developed automated computer-aided application that is in increasing use.

Object-Based Image Analysis (OBIA) is a sub-discipline of GIScience devoted to partitioning remote sensing (RS) imagery into meaningful image-objects, and assessing their characteristics through spatial, spectral and temporal scale.

Old data from remote sensing is often valuable because it may provide the only long-term data for a large extent of geography. At the same time, the data is often complex to interpret, and bulky to store. Modern systems tend to store the data digitally, often with lossless compression. The difficulty with this approach is that the data is fragile, the format may be archaic, and the data may be easy to falsify. One of the best systems for archiving data series is as computer-generated machine-readable ultrafiche, usually in typefonts such as OCR-B, or as digitized half-tone images. Ultrafiches survive well in standard libraries, with lifetimes of several centuries. They can be created, copied, filed and retrieved by automated systems. They are about as compact as archival magnetic media, and yet can be read by human beings with minimal, standardized equipment.

Generally speaking, remote sensing works on the principle of the *inverse problem*: while the object or phenomenon of interest (the state) may not be directly measured, there exists some other variable that can be detected and measured (the observation) which may be related to the object of interest through a calculation. The common analogy given to describe this is trying to determine the type of animal from its footprints. For example, while it is impossible to directly measure temperatures in the upper atmosphere, it is possible to measure the spectral emissions from a known chemical species (such as carbon dioxide) in that region. The frequency of the emissions may then be related via thermodynamics to the temperature in that region.

Data Processing Levels

To facilitate the discussion of data processing in practice, several processing "levels" were first defined in 1986 by NASA as part of its Earth Observing System and steadily adopted since then, both internally at NASA (e. g.,) and elsewhere (e. g.,); these definitions are:

Level	Description
0	Reconstructed, unprocessed instrument and payload data at full resolution, with any and all communications artifacts (e. g., synchronization frames, communications headers, duplicate data) removed.
1a	Reconstructed, unprocessed instrument data at full resolution, time-referenced, and annotated with ancillary information, including radiometric and geometric calibration coefficients and georeferencing parameters (e. g., platform ephemeris) computed and appended but not applied to the Level 0 data (or if applied, in a manner that level 0 is fully recoverable from level 1a data).
1b	Level 1a data that have been processed to sensor units (e. g., radar backscatter cross section, brightness temperature, etc.); not all instruments have Level 1b data; level 0 data is not recoverable from level 1b data.
2	Derived geophysical variables (e. g., ocean wave height, soil moisture, ice concentration) at the same resolution and location as Level 1 source data.
3	Variables mapped on uniform spacetime grid scales, usually with some completeness and consistency (e. g., missing points interpolated, complete regions mosaicked together from multiple orbits, etc.).
4	Model output or results from analyses of lower level data (i. e., variables that were not measured by the instruments but instead are derived from these measurements).

A Level 1 data record is the most fundamental (i. e., highest reversible level) data record that has significant scientific utility, and is the foundation upon which all subsequent data sets are produced. Level 2 is the first level that is directly usable for most scientific applications; its value is much greater than the lower levels. Level 2 data sets tend to be less voluminous than Level 1 data because they have been reduced temporally, spatially, or spectrally. Level 3 data sets are generally smaller than lower level data sets and thus can be dealt with without incurring a great deal of data handling overhead. These data tend to be generally more useful for many applications. The regular spatial and temporal organization of Level 3 datasets makes it feasible to readily combine data from different sources. While these processing levels are particularly suitable for typical satellite data processing pipelines, other data level vocabularies have been defined and may be appropriate for more heterogeneous workflows.

Remote Sensing with Gamma Rays

There are applications of gamma rays to mineral exploration through remote sensing. In 1972 more than two million dollars were spent on remote sensing applications with gamma rays to mineral exploration. Gamma rays are used to search for deposits of uranium. By observing radioactivity from potassium, porphyry copper deposits can be located. A high ratio of uranium to thorium has been found to be related to the presence of hydrothermal copper deposits.

Satellites

An Earth observation satellite or Earth remote sensing satellite is a satellite used or designed for Earth observation (EO) from orbit, including spy satellites and similar ones intended for non-military uses such as environmental monitoring, meteorology, cartography and others. The most common type are Earth imaging satellites, that take satellite images, analogous to aerial photographs; some EO satellites may perform remote sensing without forming pictures, such as in GNSS radio occultation. The first occurrence of satellite remote sensing can be dated to the launch of the first artificial satellite, Sputnik 1, by the Soviet Union on October 4, 1957. Sputnik 1 sent back radio signals, which scientists used to study the ionosphere. The United States Army Ballistic Missile Agency launched the first American satellite, Explorer 1, for NASA's Jet Propulsion Laboratory on January 31, 1958. The information sent back from its radiation detector led to the discovery of the Earth's Van Allen radiation belts. The TIROS-1 spacecraft, launched on April 1, 1960, as part of NASA's Television Infrared Observation Satellite (TIROS) program, sent back the first television footage of weather patterns to be taken from space.

In 2008, more than 150 Earth observation satellites were in orbit, recording data with both passive and active sensors and acquiring more than 10 terabits of data daily. By 2021, that total had grown to over 950, with the largest number of satellites operated by US-based company Planet Labs.

Most Earth observation satellites carry instruments that should be operated at a relatively low altitude. Most orbit at altitudes above 500 to 600 kilometers (310 to 370 mi). Lower orbits have significant air-drag, which makes frequent orbit reboost maneuvers necessary. The Earth observation satellites ERS-1, ERS-2 and Envisat of European Space Agency as well as the MetOp spacecraft of EUMETSAT are all operated at altitudes of about 800 km (500 mi). The Proba-1, Proba-2 and SMOS spacecraft of European Space Agency are observing the Earth from an altitude of about 700 km (430 mi). The Earth observation satellites of UAE, DubaiSat-1 & DubaiSat-2 are also placed in Low Earth orbits (LEO) orbits and providing satellite imagery of various parts of the Earth. To get global coverage with a low orbit, a polar orbit is used. A low orbit will have an orbital period of roughly 100 minutes and the Earth will rotate around its polar axis about 25° between successive orbits. The ground track moves towards the west 25° each orbit, allowing a different section of the globe to be scanned with each orbit. Most are in Sun-synchronous orbits. A geostationary orbit, at 36,000 km (22,000 mi), allows a satellite to hover over a constant spot on the earth since the orbital period at this altitude is 24 hours. This allows uninterrupted coverage of more than 1/3 of the Earth per satellite, so three satellites, spaced 120° apart, can cover the whole Earth. This type of orbit is mainly used for meteorological satellites.

Quantum Principles Underlying Remote Sensing

Remote quantum sensing leverages the core principles of quantum mechanics, such as coherence, entanglement, and spin manipulation, to detect environmental variables like magnetic fields and temperature. One of the most widely used platforms in this domain is the nitrogen-vacancy (NV) center in diamond. NV centers are point defects in the diamond lattice, composed of a nitrogen atom adjacent to a vacant carbon site. These centers exhibit spin states that are sensitive to magnetic and electric fields, strain, and temperature. Through optical and microwave techniques, these spin states can be initialized, manipulated, and read out, allowing precise measurement of external stimuli. The remote aspect is enabled by the use of light, particularly laser beams, to both interrogate and read the NV centers. Because the optical signals can travel through different media and over distances, NV-based sensors do not need to be in direct contact with the object being sensed. Additionally, quantum sensors based on trapped ions, atomic vapor cells, and superconducting circuits are also being explored for remote sensing applications due to their high sensitivity and tunability.

Remote Magnetic Field Sensing

Remote sensing of magnetic fields has critical applications in material science, medical diagnostics, navigation, and fundamental physics. Quantum sensors detect magnetic fields by observing how the presence of a field affects the energy levels or spin states of a quantum system. In the case of NV centers, the magnetic field causes a shift in the resonance frequency of the spin transitions, which can be measured optically. This technique, known as optically detected magnetic resonance (ODMR), allows high-resolution mapping of magnetic fields with spatial resolution down to the nanoscale. Remote magnetic sensing becomes particularly valuable in situations where direct access to the field source is impractical or undesirable. For instance, in biological systems, quantum sensors can remotely detect the weak magnetic fields generated by neuronal activity, muscle contractions, or molecular binding events. In geophysics, quantum magnetometers can be used to remotely survey underground mineral deposits or tectonic movements without drilling or excavation. Similarly, in security and defense applications, quantum sensors can detect magnetic anomalies associated with hidden objects or devices, enhancing surveillance and threat detection capabilities.

Advancements in Remote Temperature Sensing

Temperature sensing is another area where quantum technologies have demonstrated significant improvements in sensitivity and spatial resolution. At the quantum level, temperature affects the energy distribution and coherence properties of particles or spin systems. In NV centers, temperature changes influence the zero-field splitting parameter of the spin levels, allowing for precise thermal measurements. Since these shifts can be optically detected, NV centers act as efficient remote thermometers. Remote quantum thermometry is particularly useful in biological and medical contexts. Traditional temperature sensors often require direct contact with tissues or fluids, which can disrupt the system being measured. Quantum thermometry, by contrast, enables non-invasive measurements with minimal disturbance. This is useful for monitoring cellular or sub-cellular temperature variations, which are crucial for understanding metabolic processes, disease progression, and drug responses. Furthermore, quantum thermometers can be integrated into nanoparticles and delivered into living organisms, providing localized and dynamic thermal data that conventional sensors cannot achieve. In industrial settings, remote quantum thermometry can monitor the temperature of machinery, electronics, or chemical reactions in real time, preventing overheating and improving operational efficiency. In extreme environments, such as outer space or high-radiation zones, quantum sensors offer resilience and accuracy that classical sensors often lack, making them indispensable tools for space missions and nuclear monitoring.

Integration with Imaging and Spectroscopy

The integration of quantum sensors with imaging and spectroscopy techniques enhances the capability of remote sensing systems. For example, by combining ODMR with confocal microscopy or wide-field imaging, researchers can create high-resolution maps of magnetic fields or temperature gradients across surfaces or volumes. This allows for spatially resolved sensing, which is vital in analyzing heterogenous systems like biological tissues, composite materials, or geological samples. Spectroscopic methods can also be employed to enhance specificity and sensitivity. Quantum sensors can be tuned to respond to specific frequencies or environmental conditions, allowing selective detection of desired signals. When applied in tandem with Raman or infrared spectroscopy, quantum sensors can provide a comprehensive picture of both structural and functional characteristics of the system under observation. Such hybrid approaches expand the utility of remote sensing in fields like environmental monitoring, biomedical research, and materials characterization.

Remote Sensing in Quantum Networks and Distributed Systems

An exciting development in the field is the use of remote quantum sensors in networked or distributed configurations. By linking multiple sensors through classical or quantum communication channels, it is possible to monitor large areas or complex systems with coordinated precision. These sensor networks can be deployed for climate monitoring, seismic activity detection, or security surveillance. Quantum entanglement, in particular, allows the creation of correlated sensor arrays where information from one sensor improves the accuracy of others. Such distributed sensing systems benefit from redundancy, real-time data sharing, and fault tolerance. They are also well-suited for applications requiring spatial-temporal resolution, such as tracking the movement of magnetic fields over time or observing temperature fluctuations across different zones. As quantum communication technologies advance, the development of quantum sensor networks is expected to play a major role in future smart infrastructure and global monitoring systems.

Challenges and Engineering Considerations

While the benefits of remote quantum sensing are clear, several challenges need to be addressed to realize their full potential. One major issue is signal degradation over distance, particularly in optical systems where scattering and absorption can reduce signal quality. Developing robust and efficient optical components, such as fiber-optic links and laser sources, is essential for maintaining sensor fidelity during remote operations. Additionally, environmental noise, including thermal fluctuations and electromagnetic interference, can affect the coherence of quantum systems. Advanced error correction techniques, shielding materials, and sensor calibration methods are needed to mitigate these effects. Miniaturization and portability are also important goals for practical deployment. Integrating quantum sensors into compact, battery-powered devices suitable for field use remains a technological challenge. Recent advances in diamond fabrication, chip-scale atomic devices, and nanophotonics are making progress in this direction, enabling smaller and more energy-efficient sensors. Furthermore, translating laboratory demonstrations into scalable, cost-effective technologies suitable for mass production and commercial use will require interdisciplinary collaboration among physicists, engineers, and manufacturers. Software development, including algorithms for data interpretation and sensor control, also plays a crucial role in transforming raw quantum signals into actionable information.

Future Directions and Impact

The future of remote sensing using quantum technologies is exceptionally promising. Continued improvements in quantum materials, optical control, and data processing will lead to more sensitive, reliable, and user-friendly devices. As these technologies mature, they are expected to find widespread applications in healthcare, environmental science, space exploration, national security, and beyond. Quantum-enhanced remote sensing could transform our ability to detect early signs of diseases, monitor critical infrastructure, observe environmental changes, and study the universe. By allowing high-resolution, non-invasive, and real-time measurements, quantum sensors will redefine standards for precision and efficiency in a wide array of disciplines.

5.4.2 Sensing in Extreme Conditions (e.g., Deep Ocean, Space)

Sensing in extreme conditions refers to the use of advanced sensor technologies to detect and measure physical parameters in environments that are characterized by harsh, challenging, or otherwise inhospitable conditions. Such environments include the deep ocean, outer space, polar regions, high-radiation zones, and high-pressure or high-temperature industrial settings. These environments often pose unique challenges for conventional sensing technologies, including issues related to signal degradation, equipment durability, environmental noise, and limitations in power supply and data transmission. As human exploration and industrial operations extend further into such domains, the need for robust, precise, and resilient sensing technologies becomes more urgent.

Sensing in extreme conditions is a frontier of technology that enables humanity to extend its reach into the most challenging and uncharted environments on Earth and beyond. The unique demands of deep ocean and space exploration require sensors that are not only precise but also resilient, autonomous, and efficient. Quantum sensing technologies are poised to play a transformative role in this field, offering unprecedented levels of sensitivity and robustness. Looking ahead, continued progress in material science, miniaturization, data processing, and quantum engineering will drive the development of next-generation sensors capable of enduring even harsher conditions. These advances will open new avenues for scientific research, environmental monitoring, and industrial innovation, further expanding our understanding of the universe and our capacity to operate within it.

Challenges of Deep Ocean Sensing

The deep ocean represents one of the most extreme and least understood environments on Earth. Characterized by immense pressure, near-freezing temperatures, total darkness, and corrosive salinity, it challenges the operational capacity of many conventional sensors. At depths exceeding several kilometers, the pressure can rise to hundreds of atmospheres, requiring sensors and their enclosures to be highly pressure-resistant. The lack of natural light in these regions eliminates the feasibility of optical sensing without artificial illumination, while the corrosive nature of saltwater can degrade sensor materials over time. Furthermore, deploying and retrieving sensors in such remote environments can be costly and complex. Autonomous underwater vehicles (AUVs) and remotely operated vehicles (ROVs) are often used to carry sensors to the desired depths. These sensors may be tasked with measuring temperature, salinity, pressure, magnetic fields, and chemical compositions, contributing valuable data for oceanographic research, undersea resource exploration, and environmental monitoring. Quantum-enhanced sensors, which rely on phenomena such as atomic coherence or spin resonance, are being explored to provide higher sensitivity and reliability in these difficult environments.

Sensor Requirements for Space Missions

Outer space presents another domain of extreme conditions, including vacuum, microgravity, wide temperature fluctuations, intense radiation, and prolonged periods without human intervention. Sensing technologies in space are vital for a wide range of missions, including planetary exploration, satellite operations, and astronomical observations. Unlike the deep ocean, where pressure is extremely high, space is a near-perfect vacuum, requiring sensors to be completely sealed and shielded from the absence of atmospheric pressure. The temperature in space can vary dramatically depending on exposure to solar radiation, with extremes ranging from hundreds of degrees above zero in sunlight to cryogenic temperatures in the shade. This necessitates the use of sensors and electronics that can maintain performance despite extreme thermal cycling. Additionally, space radiation—composed of high-energy particles from the sun and cosmic sources—can damage sensitive electronics and alter sensor readings. Hence, radiation-hardened sensors with built-in redundancy and fault-tolerant designs are essential. Quantum technologies are increasingly being considered for space missions due to their potential for extremely sensitive measurements of gravitational fields, magnetic fields, and inertial motion. For example, quantum accelerometers and gyroscopes can help spacecraft navigate autonomously without relying on GPS, which is not available beyond Earth's orbit. Quantum magnetometers and atomic clocks, similarly, offer unparalleled precision for scientific measurements and mission timing.

Materials and Engineering Considerations

To function reliably in extreme environments, sensors must be built from materials that can withstand the physical stresses they will encounter. In the deep ocean, sensor housings are typically constructed from titanium, ceramics, or specialized polymers that resist corrosion and endure high pressure. In space, lightweight materials such as aluminum alloys, carbon composites, and radiation-resistant coatings are often employed to reduce launch mass while ensuring durability. Thermal management is a critical concern in both domains. In space, sensors are often equipped with radiative heat shields or thermal blankets to maintain a stable internal temperature. In the ocean, sensors may rely on insulation and temperature compensation algorithms to ensure accuracy despite the cold surroundings. The design of the sensor circuitry also plays a major role. Low-power consumption, high signal-to-noise ratio, and minimal drift over time are essential features for long-term, unattended deployments.

Remote Operation and Data Transmission

Another critical aspect of sensing in extreme conditions is the method of communication and data transmission. In the deep ocean, radio frequency communication is largely ineffective due to high signal attenuation in seawater. Acoustic communication, while viable, is limited in bandwidth and susceptible to noise from marine life and underwater activity. As a result, data may need to be stored onboard the sensor platform and retrieved later or transmitted intermittently when conditions permit. In space, communication is typically achieved using radio or laser signals, but bandwidth and latency are significant constraints, particularly for deep-space missions. Sensors must be capable of autonomous operation, making decisions about what data to record or transmit. This necessitates the integration of smart sensing systems with embedded computing and artificial intelligence algorithms that can filter, compress, and prioritize data before transmission.

Quantum Sensing for Extreme Environments

Quantum sensing technologies offer several advantages for operation in extreme conditions. These sensors exploit quantum properties such as entanglement, superposition, and tunneling to detect physical quantities with extraordinary precision. One of the most promising features of quantum sensors is their high sensitivity to minute environmental changes, allowing them to function effectively where classical sensors struggle. In the deep ocean, quantum magnetometers can be used to detect variations in Earth's magnetic field that indicate geological activity or the presence of undersea resources. These sensors can be deployed on autonomous platforms to perform long-duration missions. In space, quantum interferometers can detect gravitational waves and other minute distortions in spacetime, contributing to our understanding of fundamental physics. Quantum clocks provide extremely accurate timing references that are critical for navigation and synchronization across space-based systems. Furthermore, advances in diamond-based sensors, particularly those utilizing nitrogen-vacancy centers, are enabling remote sensing of temperature and magnetic fields with high spatial resolution. These sensors are robust to environmental stress and can operate under ambient or cryogenic conditions, making them ideal for both deep ocean and space applications.

Applications in Scientific Research and Industry

The ability to sense accurately in extreme environments opens up numerous possibilities for scientific discovery and industrial applications. In oceanography, sensors contribute to understanding climate change by monitoring deep-sea currents, temperature gradients, and carbon storage. They also support marine biology by tracking animal behavior and habitat conditions in otherwise inaccessible regions. In space, sensors are central to missions that aim to explore other planets, search for life, and study cosmological phenomena. They are also crucial for maintaining the health and safety of spacecraft and satellites, detecting faults, managing onboard systems, and ensuring mission success. Commercially, companies are investing in deep-sea mining and offshore energy production, where robust sensors are needed for navigation, drilling, and environmental compliance. In the aerospace sector, quantum-enhanced sensors are expected to improve spacecraft guidance systems, enable precision landings, and enhance Earth observation from orbit.

5.5 QUANTUM SENSING IN FUNDAMENTAL PHYSICS AND COSMOLOGY

5.5.1 Precision Measurement of Fundamental Constants

The pursuit of precise measurements of fundamental constants has long been at the core of physics. These constants—such as the speed of light, Planck's constant, the fine-structure constant, and the gravitational constant—serve as the foundation upon which physical theories are built. The ability to measure them with greater precision not only strengthens our understanding of known physics but also opens new pathways to discovering unknown phenomena. In recent decades, quantum sensing has emerged as a transformative approach in the domain of precision measurement. Leveraging the inherent sensitivity of quantum systems to external influences, quantum sensing offers unmatched accuracy in probing the values of these constants. This field is now making significant contributions to both fundamental physics and cosmology. Quantum sensing represents a significant advance in the field of precision measurement. By utilizing the remarkable properties of quantum systems, researchers are now able to measure fundamental constants with levels of accuracy that were previously unattainable. These efforts not only strengthen the empirical foundation of modern physics but also create opportunities to explore new realms of fundamental science. From redefining SI units to probing the validity of the Standard Model and understanding the early universe, quantum sensors are becoming indispensable tools in the physicist's arsenal. As the technology continues to evolve, its role in unraveling the deepest mysteries of the universe is only expected to grow, heralding a new era in fundamental physics and cosmology.

Role of Quantum Systems in Enhancing Measurement Accuracy

Quantum systems possess several properties that make them ideal candidates for ultra-precise measurements. Coherence, entanglement, and superposition enable quantum sensors to outperform their classical counterparts in terms of sensitivity and resolution. One of the most prominent examples is the use of atomic interferometry to measure gravitational effects or variations in inertial forces. By manipulating atomic wavefunctions and allowing them to interfere, it becomes possible to detect incredibly subtle differences in phase that reflect the influence of gravitational or electromagnetic fields. Atomic clocks, which use the frequency of electromagnetic transitions in atoms such as cesium or ytterbium, have achieved extraordinary precision in timekeeping. The measurement of time and frequency with such clocks is directly related to the precision of other constants, like Planck's constant or the Rydberg constant. Improvements in clock accuracy have enabled fundamental tests of physical laws, such as time dilation in special and general relativity, and they also provide a reference for redefining SI units.

Determining Planck's Constant with Quantum Sensors

Planck's constant is a cornerstone of quantum mechanics, linking the energy of photons to their frequency. Precise knowledge of this constant is essential for understanding the behavior of quantum systems and for determining other constants like the elementary charge. Quantum sensing methods have played a vital role in measuring Planck's constant with unprecedented accuracy. The development of the Kibble balance, formerly known as the watt balance, exemplifies how quantum principles can be applied to macroscopic measurements. This instrument balances mechanical power against electromagnetic power using measurements that involve the Josephson and quantum Hall effects—both of which rely on quantum phenomena. The Josephson effect links voltage to frequency via Planck's constant, while the quantum Hall effect ties resistance to fundamental constants. Together, these effects allow researchers to determine Planck's constant through purely quantum means, ultimately contributing to the redefinition of the kilogram in 2019.

Fine-Structure Constant and Quantum Electrodynamics

Another fundamental parameter of great importance is the fine-structure constant, which characterizes the strength of electromagnetic interactions between charged particles. It is dimensionless and appears in many equations in quantum electrodynamics (QED). Measuring it with high precision allows physicists to test the predictions of QED and to look for discrepancies that might indicate new physics. Quantum sensors, especially those based on atomic spectroscopy and interferometry, have enabled increasingly accurate measurements of the fine-structure constant. Techniques involving laser-cooled atoms and interferometric setups permit the determination of the recoil velocity of atoms when they absorb or emit photons. This recoil is directly related to the fine-structure constant. By improving the control and measurement of such quantum interactions, researchers can refine the value of the constant and examine the internal consistency of quantum theories.

Gravitational Constant and Quantum Interferometry

The gravitational constant, denoted as G, is notoriously difficult to measure with high precision due to the weakness of the gravitational force compared to other fundamental forces. Traditional methods suffer from various sources of systematic error, including environmental noise and imperfections in mechanical systems. Quantum interferometry, using atoms or Bose-Einstein condensates (BECs), presents a new approach for measuring G with potentially greater accuracy. In these experiments, atoms are cooled to near absolute zero and placed into a superposition of momentum states. As they fall under gravity, the interference pattern created by the recombination of atomic wavefunctions encodes information about the gravitational field. By carefully controlling the experimental conditions and employing quantum-enhanced detection techniques, it is possible to extract a precise value for G. While still in developmental stages, such experiments suggest that quantum sensors could eventually provide a more reliable and accurate measurement of the gravitational constant.

Implications for Cosmology and the Early Universe

Beyond the Earth-based measurement of constants, quantum sensing has implications for cosmology and our understanding of the early universe. Several cosmological models depend on precise knowledge of fundamental constants to describe the evolution of the universe, the behavior of dark matter and dark energy, and the unification of fundamental forces. Atomic clocks and optical lattice clocks are now being deployed in space-based missions to test the constancy of these values across vast distances and timescales. For instance, if fundamental constants like the fine-structure constant or the electron-proton mass ratio varied across the universe, it would have profound implications for our models of cosmology. Quantum sensors aboard satellites can compare time and frequency references in different gravitational potentials, providing evidence for or against such variations. In addition, quantum magnetometers and gravimeters may be used to map the gravitational anomalies or detect relic signals from the early universe. These could include gravitational waves or other distortions that originated during cosmic inflation. The sensitivity of quantum sensors to such minuscule effects offers the tantalizing possibility of detecting features that would otherwise be invisible to classical instruments.

Testing the Boundaries of the Standard Model

Precision measurements facilitated by quantum sensing are also vital in testing the limits of the Standard Model of particle physics. Any deviation from expected values in the constants might hint at physics beyond the Standard Model, such as supersymmetry, extra dimensions, or unknown particles. For example, discrepancies in the magnetic moment of the muon or in the behavior of neutrinos are currently under intense investigation, and future quantum sensing technologies could provide critical data to support or refute these findings. In parallel, quantum sensors are being developed for detecting weak forces and particles, such as axions or hypothetical dark matter particles, which might subtly influence atomic behavior. These detectors can measure ultra-weak fields or rare events by observing changes in coherence, phase, or spin alignment of quantum states, pushing the boundaries of what can be observed.

5.5.2 Testing Fundamental Theories in Physics

The quest to understand the universe has always revolved around developing, refining, and testing fundamental theories that explain the behavior of nature at the deepest levels. From Newtonian mechanics to quantum mechanics and Einstein's theory of general relativity, these frameworks have offered profound insights into the workings of matter, energy, space, and time. However, as experimental techniques become increasingly precise and theoretical predictions delve into more complex domains, the need to rigorously test these theories grows ever more essential. Testing fundamental theories in physics is not only a matter of validation but also a powerful approach to uncovering potential inconsistencies or limitations that could lead to revolutionary discoveries. The testing of fundamental theories in physics is an ever-evolving process driven by the interplay between theoretical innovation and experimental ingenuity. As our measurement techniques become increasingly refined, we gain the power to probe nature at levels previously unimaginable. Whether confirming the robustness of established theories or revealing subtle cracks that hint at deeper truths, these tests form the cornerstone of scientific progress. In this dynamic landscape, quantum sensing, gravitational wave astronomy, particle accelerators, and cosmological observations all play vital roles in exploring the frontiers of knowledge. The continued pursuit of rigorous tests not only strengthens our existing frameworks but also prepares us for the paradigm shifts that may define the future of physics.

Foundational Theories and Their Experimental Verifications

At the heart of modern physics lie several foundational theories, including classical mechanics, quantum mechanics, electromagnetism, thermodynamics, the Standard Model of particle physics, and general relativity. These theories have been extensively verified through countless experiments and have stood the test of time under a wide range of conditions. Yet, despite their predictive success, there remain key areas where these theories appear to be incomplete or mutually incompatible. For example, while quantum mechanics and general relativity are each remarkably accurate in their respective domains, they resist unification into a single coherent framework. To test these foundational theories, physicists rely on precision experiments that aim to detect discrepancies between theoretical predictions and observed phenomena. These tests span both the smallest scales—where quantum effects dominate—and the largest scales—where gravitational dynamics come into play. Whether through particle collisions in high-energy accelerators or observations of distant cosmic events, experimental tests serve as the ultimate arbiter of scientific truth.

Quantum Mechanics: Probing the Foundations

Quantum mechanics, formulated in the early 20th century, has proven to be one of the most accurate and far-reaching theories in physics. It has successfully explained atomic structure, chemical bonding, and the behavior of subatomic particles. However, it also presents counterintuitive features such as superposition, entanglement, and wavefunction collapse that challenge our classical intuitions. One area of active testing involves quantum entanglement, a phenomenon where particles share a quantum state in such a way that measurements on one instantaneously affect the other, regardless of the distance separating them. Bell's theorem and subsequent experiments have demonstrated that local hidden variable theories cannot explain these correlations, affirming the non-local nature of quantum mechanics. Modern tests of Bell inequalities, especially those that close loopholes related to detector inefficiencies and communication delays, continue to strengthen the case for quantum theory while probing the boundaries of realism and locality. Other experimental efforts involve quantum decoherence, quantum tunneling, and quantum measurement. These tests not only validate the predictions of quantum mechanics but also serve to explore new interpretations and potential modifications of the theory. In particular, researchers are interested in whether quantum mechanics remains valid at macroscopic scales, a question being addressed through experiments involving superconducting circuits, optomechanical systems, and Bose-Einstein condensates.

General Relativity and Gravitational Testing

Einstein's theory of general relativity revolutionized our understanding of gravity by describing it as the curvature of spacetime caused by mass and energy. This elegant theory has passed numerous observational tests, including the bending of starlight by gravity, the precession of Mercury's orbit, and the precise timing of signals from pulsars in binary systems. More recently, the detection of gravitational waves by LIGO and Virgo has opened a new era in testing general relativity. These ripples in spacetime, generated by violent astrophysical events like black hole mergers, carry information about the extreme gravitational environments where the theory can be critically examined. So far, observations of gravitational waveforms match the predictions of general relativity with remarkable accuracy, but future detections may reveal subtle deviations or unknown phenomena. Space-based missions such as the Gravity Probe B, the MICROSCOPE satellite, and ongoing satellite laser ranging experiments are also designed to test the equivalence principle, geodetic effects, and the universality of free fall—cornerstones of general relativity. These tests seek to uncover any anomalies that might suggest the need for alternative theories or quantum modifications of gravity.

The Standard Model and High-Energy Physics

The Standard Model of particle physics describes the electromagnetic, weak, and strong nuclear interactions using quantum field theories. It has accurately predicted the existence and behavior of particles such as quarks, leptons, bosons, and the Higgs boson. However, it leaves several questions unanswered, including the nature of dark matter, the imbalance between matter and antimatter, and the origin of neutrino masses. Testing the Standard Model involves probing the structure of matter at ever higher energies and finer resolutions. Facilities like the Large Hadron Collider (LHC) at CERN are central to this endeavor. By colliding protons at near-light speeds, researchers can create and study particles that existed only in the early universe. Experiments at the LHC confirmed the existence of the Higgs boson in 2012, a pivotal moment in validating the Standard Model. Nevertheless, physicists continue to look for signs of new physics beyond the Standard Model. These include the possible existence of supersymmetry, extra spatial dimensions, or particles that could account for dark matter. Precision measurements of rare decays, magnetic moments, and symmetry violations are key methods for identifying discrepancies that may point toward new theories.

Cosmology and Tests of Large-Scale Physics

Cosmology provides a unique laboratory for testing fundamental theories across vast distances and time scales. Observations of the cosmic microwave background radiation, the distribution of galaxies, and the expansion rate of the universe offer valuable data for testing both general relativity and quantum field theory in extreme conditions. One major focus is the investigation of dark energy, the mysterious force driving the accelerated expansion of the universe. The standard cosmological model, known as Lambda-CDM, incorporates dark energy as a cosmological constant, but its origin and precise nature remain elusive. Alternative theories, including modifications of gravity or the existence of dynamic scalar fields, are being explored to explain the observations. Another key challenge involves the apparent discrepancies in the measured value of the Hubble constant. Different observational techniques—one based on early-universe data (like the cosmic microwave background) and another based on supernova distances—yield slightly different results. This tension may suggest the need for revisions in our understanding of cosmology and physics at large scales.

Testing Symmetries and Conservation Laws

Symmetries and conservation laws lie at the heart of theoretical physics. The invariance of physical laws under translations, rotations, and boosts gives rise to conserved quantities like momentum and energy. Testing these symmetries, particularly charge-parity (CP) symmetry and time-reversal symmetry, is crucial for understanding the limitations of current theories. Violations of CP symmetry, for instance, have been observed in the behavior of certain mesons, and they are essential for explaining the dominance of matter over antimatter. However, the degree of CP violation in the Standard Model appears insufficient to account for this asymmetry, prompting the search for additional sources of symmetry breaking. Precision experiments, including those with ultracold neutrons, atomic clocks, and electric dipole moment measurements, seek to uncover violations of time-reversal symmetry and Lorentz invariance. Discovering such violations would have profound implications for fundamental physics, potentially pointing to new fields or interactions.

CHAPTER 6

Key Quantum Sensing Technologies

6.1 NITROGEN-VACANCY (NV) CENTERS IN DIAMOND

Nitrogen-vacancy (NV) centers in diamond represent one of the most remarkable quantum systems in solid-state physics, bridging the realms of condensed matter physics, quantum optics, and quantum sensing. These centers are point defects in the diamond lattice, consisting of a nitrogen atom substituting for a carbon atom adjacent to a lattice vacancy. Their unique electronic and spin properties make NV centers highly valuable for applications in quantum computing, high-resolution sensing, and fundamental physics research. The diamond lattice provides a robust, chemically inert, and optically transparent host, allowing NV centers to function at room temperature and under ambient conditions, which significantly expands their utility in practical technologies. Nitrogen-vacancy centers in diamond have emerged as one of the most versatile and promising tools in quantum science and technology.

Their ability to function at room temperature, combined with their unique spin and optical properties, has enabled a wide range of applications spanning quantum sensing, bioimaging, magnetometry, and quantum information processing. Continued progress in material engineering, device integration, and quantum control promises to unlock even greater potential for NV centers in the coming decades. As research in this field evolves, NV centers are expected to play a central role in the development of quantum-enabled technologies that address pressing challenges in science, medicine, and industry.

6.1.1 Structure and Electronic Properties

The NV center in diamond occurs in two charge states: neutral (NV^0) and negatively charged (NV^-), with the latter being the most relevant for quantum applications. The NV^- center consists of a nitrogen atom next to a vacancy that captures an extra electron, resulting in a total of six electrons in the defect system. These electrons give rise to spin-triplet ground and excited states, which are optically active and possess spin-dependent fluorescence. The ground state, in particular, exhibits a zero-field splitting between the ms=0m_s = 0ms=0 and ms=±1m_s = \pm1ms=±1 spin sublevels, typically around 2.87 GHz. This spin structure allows for precise manipulation using microwave fields and optical detection via fluorescence changes, forming the foundation of optically detected magnetic resonance (ODMR).

6.1.2 Quantum Coherence and Spin Control

One of the most attractive features of NV centers is their exceptional spin coherence, which persists even at room temperature. The coherence time, or T2T_2T2, can reach several milliseconds in ultra-pure diamond samples with low concentrations of magnetic impurities such as 13^{13}13C. This long coherence time makes NV centers suitable for various quantum applications, including quantum memory, quantum computation, and precision sensing.

The spin state of the NV center can be initialized using green laser light, manipulated with radio-frequency or microwave pulses, and read out optically through spin-dependent photoluminescence. The ms=0m_s = 0ms=0 state emits more photons than the ms=±1m_s = \pm1ms=±1 states, enabling high-fidelity single-shot readout under appropriate conditions.

6.1.3 Optical Readout and ODMR Techniques

Optically detected magnetic resonance is central to most applications of NV centers. It combines microwave excitation with optical pumping and readout to detect spin transitions in the NV center. When a resonant microwave field is applied, the NV spin state flips between ms=0m_s = 0ms=0 and ms=±1m_s = \pm1ms=±1, leading to a measurable drop in fluorescence intensity. This technique enables high-sensitivity measurements of magnetic fields, electric fields, and temperature changes. ODMR works effectively at room temperature and in various environments, making it an incredibly versatile and accessible quantum sensing method. The narrow linewidth of the resonance and the spin's long coherence time allow for precise frequency measurements, which translate into accurate sensing capabilities.

6.1.4 Magnetic Sensing with NV Centers

One of the most well-developed applications of NV centers is in magnetometry. Due to the Zeeman effect, external magnetic fields shift the resonance frequencies of the NV spin states. This shift can be measured with high precision, allowing NV centers to act as highly sensitive, nanoscale magnetometers. These properties are particularly useful for studying magnetic phenomena in condensed matter systems, including imaging magnetic vortices in superconductors, monitoring current flow in nanoelectronic circuits, and investigating biological systems with magnetically tagged molecules. NV-based magnetometers can operate in wide-field imaging mode, enabling two-dimensional mapping of magnetic fields across extended surfaces, or in scanning-probe mode, achieving sub-nanometer spatial resolution.

6.1.5 Electric Field and Temperature Sensing

In addition to magnetic field detection, NV centers are sensitive to electric fields and temperature variations. Changes in electric fields affect the NV energy levels through the Stark effect, while temperature shifts influence the zero-field splitting between the spin sublevels due to lattice vibrations and strain. By measuring these changes, NV centers can be employed as nano-thermometers and electrostatic field sensors. These capabilities are particularly valuable in biology and materials science, where local temperature and charge distribution play crucial roles. For instance, NV-based thermometry has been used to monitor thermal gradients in living cells and to study heat dissipation in electronic devices at the nanoscale.

6.1.6 Biological Applications and Imaging

The biocompatibility and photostability of diamond make NV centers particularly suitable for biological applications. Nanodiamonds containing NV centers can be functionalized with biomolecules and internalized by cells without causing toxicity or significant cellular disruption. Once inside, they can serve as fluorescent markers or local sensors for temperature and magnetic field changes. NV-based imaging is advantageous over traditional fluorescent dyes because NV centers do not bleach or blink and can be excited with low-power lasers, reducing photodamage to biological samples. Applications range from tracking intracellular processes to detecting magnetic nanoparticles used in targeted drug delivery and diagnostics.

6.1.7 Quantum Information Processing

NV centers also play a significant role in the field of quantum information science. Their electron spins can function as quantum bits (qubits) that store and manipulate quantum information. Moreover, nearby nuclear spins, such as those from the nitrogen nucleus or adjacent 13^{13}13C atoms, can serve as auxiliary qubits, forming a small quantum register. Quantum gates can be implemented using microwave pulses and hyperfine interactions, allowing for basic quantum algorithms and error correction schemes. Additionally, the spin-photon interface offered by NV centers enables long-distance quantum communication through entanglement and photon-mediated quantum networking.

6.1.8 Material Science and Quantum Engineering

The performance of NV centers is closely linked to the quality of the diamond host material. High-purity synthetic diamonds grown by chemical vapor deposition (CVD) or high-pressure high-temperature (HPHT) methods allow for controlled doping and reduced defect densities, leading to improved coherence times and reduced background noise. Advanced fabrication techniques, including ion implantation and annealing, are used to create NV centers at specific depths and locations, which is crucial for integrated quantum devices and scanning probe applications. Recent efforts in quantum engineering aim to integrate NV centers with photonic structures such as waveguides and resonators, enhancing photon collection efficiency and enabling scalable quantum networks.

6.1.9 Challenges and Future Directions

Despite their many advantages, NV centers face several technical challenges that limit their performance in certain applications. These include photon collection inefficiencies, limited spin readout fidelity, and environmental noise from nearby defects or nuclear spins. Addressing these issues requires advances in diamond fabrication, surface chemistry, and quantum control techniques. On the theoretical front, efforts continue to better understand spin dynamics, decoherence mechanisms, and optimal measurement protocols. Future directions include the integration of NV centers with complementary quantum systems, such as superconducting qubits or trapped ions, to create hybrid quantum platforms with enhanced capabilities.

6.2 ATOMIC CLOCKS AND QUANTUM TIMING

Atomic clocks represent one of the most precise and transformative achievements in modern science, fundamentally redefining our understanding and measurement of time. At the heart of atomic clocks lies the principle of quantum timing, which exploits the intrinsic properties of atoms and their transitions between energy levels to establish an ultra-precise frequency standard. Unlike classical mechanical clocks that rely on the oscillation of pendulums or quartz crystals, atomic clocks utilize the natural oscillations of electrons within atoms when they transition between quantum states. These transitions occur at extremely consistent frequencies, enabling timekeeping with exceptional stability and accuracy. Quantum timing, therefore, stands at the intersection of atomic physics, quantum mechanics, and precision engineering, forming the foundation for many contemporary technologies, including GPS, telecommunications, and fundamental tests of physical laws. Atomic clocks, grounded in the principles of quantum timing, represent a pinnacle of precision in modern science. From the cesium-based microwave standards of the 20th century to today's optical and quantum-enhanced systems, these clocks have profoundly influenced technology, science, and society. Their unmatched accuracy enables global positioning, synchronizes communication networks, and facilitates cutting-edge research in physics and cosmology. As advancements in quantum control, entanglement, and miniaturization continue, atomic clocks are poised to become even more central to our understanding of time and the universe. Through their continued development, humanity is not only keeping better time but is also expanding the frontiers of knowledge and innovation.

6.2.1 The Quantum Basis of Atomic Timekeeping

The foundation of atomic timekeeping lies in the regularity of atomic transitions, which are governed by the laws of quantum mechanics. When an atom transitions between two energy states, it absorbs or emits electromagnetic radiation at a frequency that is uniquely determined by the difference in energy between those states. This frequency is constant and universal for a given type of atom, making it an ideal reference for timekeeping. For example, in cesium-133, the atom used in the first practical atomic clock, the transition between two hyperfine ground states occurs at exactly 9,192,631,770 cycles per second. This number is used to define the SI second, making the second one of the only units in the International System of Units that is based on a fundamental physical constant rather than a physical artifact or object.

6.2.2 Evolution from Cesium Clocks to Optical Clocks

The earliest atomic clocks were based on cesium atoms and operated using microwave frequencies. These clocks, known as cesium beam clocks, have provided reliable time standards for decades and continue to serve as the basis for International Atomic Time (TAI). However, advances in laser technology and atomic physics have given rise to a new generation of atomic clocks that operate in the optical domain. Optical atomic clocks use atoms such as strontium, ytterbium, and aluminum ions, which have transitions at much higher frequencies than microwaves. These higher frequencies allow for finer resolution in time measurement, significantly improving both precision and stability. Optical clocks can achieve accuracies better than one part in 10^18, meaning they would lose or gain less than a second over the entire age of the universe.

6.2.3 Laser Cooling and Trapping in Atomic Clocks

A critical advancement in the development of high-precision atomic clocks is the use of laser cooling and trapping techniques. By cooling atoms to near absolute zero, their thermal motion is significantly reduced, which minimizes Doppler broadening and other sources of frequency instability. Optical lattices and ion traps further confine the atoms, allowing for more precise control over their environment and interactions. In optical lattice clocks, for instance, neutral atoms are held in place by a standing wave of laser light, allowing many atoms to be interrogated simultaneously without significant movement. This leads to an increase in signal strength and a reduction in statistical noise, contributing to the improved performance of modern atomic clocks.

6.2.4 Quantum Entanglement and Enhanced Clock Performance

Recent developments in quantum information science have introduced the possibility of using quantum entanglement to further enhance atomic clock performance. Entangled states allow atoms to share quantum correlations that can be harnessed to reduce measurement uncertainty beyond the standard quantum limit, achieving what is known as the Heisenberg limit. By creating entangled ensembles of atoms, researchers can improve the signal-to-noise ratio in frequency measurements, leading to more precise timekeeping. Although the practical implementation of entangled atomic clocks is still in its early stages, the theoretical foundation suggests substantial gains in both stability and sensitivity, paving the way for next-generation time standards.

6.2.5 Applications of Atomic Clocks in Modern Technology

Atomic clocks have become indispensable in a wide array of technological systems that require high-precision timing. One of the most prominent applications is in global navigation satellite systems (GNSS) such as GPS, GLONASS, Galileo, and BeiDou. Each satellite in these systems carries atomic clocks that broadcast time signals, which receivers on Earth use to triangulate their position with meter-level accuracy. Without atomic clock precision, GPS systems would rapidly drift, rendering them ineffective for navigation. In telecommunications, atomic clocks ensure synchronization across data networks, enabling seamless communication and data transfer. They also play a vital role in high-frequency trading, scientific experiments requiring synchronized observations, and secure time stamping in cybersecurity.

6.2.6 Atomic Clocks and Fundamental Physics

Beyond practical applications, atomic clocks are powerful tools for testing the fundamental laws of physics. Due to their extreme sensitivity to external influences, they can detect minute variations in physical constants, search for evidence of dark matter, and test theories of relativity. For instance, gravitational time dilation, a prediction of Einstein's general relativity, has been confirmed using atomic clocks placed at different altitudes on Earth, which experience slightly different gravitational potentials. Moreover, comparisons between different types of atomic clocks can reveal potential variations in fundamental constants such as the fine-structure constant or the electron-to-proton mass ratio. These comparisons serve as critical tests for theories that extend beyond the Standard Model of particle physics.

6.2.7 Towards a Global Quantum Time Network

With the emergence of optical clocks and quantum-enhanced timing systems, there is growing interest in creating a global quantum time network. Such a network would link atomic clocks in different locations through quantum communication channels or optical fiber links, allowing for real-time synchronization with unprecedented accuracy. This could revolutionize timekeeping, enabling applications that require ultra-precise synchronization across vast distances, such as interferometric telescope arrays and distributed quantum computing networks. Additionally, a global quantum time network would provide a resilient and tamper-proof infrastructure for critical systems such as financial markets, power grids, and defense technologies.

6.2.8 Challenges in Atomic Clock Development

Despite their impressive performance, atomic clocks face several technical and practical challenges. Environmental noise, magnetic field fluctuations, and thermal shifts can introduce errors that must be carefully compensated. The complexity and cost of building and maintaining optical atomic clocks have limited their widespread deployment outside specialized laboratories. Furthermore, the miniaturization of atomic clocks for portable and space-based applications remains an ongoing area of research. Chip-scale atomic clocks (CSACs) have made significant progress in this direction, offering reasonable precision in a compact form factor suitable for mobile and embedded systems, although they do not yet rival the performance of their larger counterparts.

6.2.9 The Future of Timekeeping

The field of quantum timing and atomic clocks continues to evolve, driven by scientific curiosity and technological demands. As optical clocks become more robust and scalable, they are expected to redefine the second in the International System of Units, replacing the cesium-based definition. Future research aims to integrate quantum technologies, artificial intelligence, and new materials to enhance clock stability, portability, and resilience. The exploration of new atomic and molecular species, along with exotic states of matter such as Bose-Einstein condensates, may yield even better time standards. Ultimately, atomic clocks are not merely tools for keeping time but are instruments that probe the fabric of reality, offering deeper insights into the universe and enabling the next wave of technological innovation.

6.3 SUPERCONDUCTING SENSORS AND SQUIDS

Superconducting sensors, particularly Superconducting Quantum Interference Devices (SQUIDs), have emerged as some of the most sensitive tools available for measuring magnetic fields and other physical quantities. These devices exploit the unique properties of superconductors, materials that exhibit zero electrical resistance and the expulsion of magnetic fields below a certain critical temperature. SQUIDs utilize quantum interference effects in superconducting loops containing Josephson junctions to detect extremely small changes in magnetic flux. Since their invention in the 1960s, SQUIDs have found wide-ranging applications in fields such as medicine, geophysics, materials science, and fundamental physics.

Their unmatched sensitivity makes them ideal for tasks where the detection of minute magnetic signals is crucial, often reaching sensitivity levels down to a few femtoteslas. The convergence of superconductivity, quantum mechanics, and precision measurement technology makes superconducting sensors and SQUIDs vital components in the modern scientific and technological landscape. Superconducting sensors and SQUIDs represent a convergence of quantum mechanics, superconductivity, and precision engineering that has revolutionized the way we measure magnetic fields and related physical phenomena. Their extraordinary sensitivity, grounded in the quantum behavior of Josephson junctions and superconducting loops, has enabled breakthroughs in medicine, physics, environmental science, and beyond. While challenges related to cryogenics and noise persist, advances in materials and technology continue to expand their capabilities and accessibility. As the demand for ultra-sensitive detection grows across disciplines, SQUIDs and related superconducting sensors are poised to remain at the forefront of scientific innovation and technological progress.

6.3.1 Principles of Superconductivity and Quantum Interference

Superconductivity provides the foundational physical phenomena that enable the operation of SQUIDs. When a material becomes superconducting, it transitions into a quantum state characterized by the coherent behavior of paired electrons known as Cooper pairs. This quantum coherence allows for unique effects such as the quantization of magnetic flux and the Josephson effect, both of which are critical for the functioning of SQUIDs. The Josephson effect occurs when two superconductors are separated by a thin insulating layer, forming a Josephson junction. In such a configuration, a supercurrent can tunnel through the barrier without any applied voltage, and this current depends sensitively on the phase difference of the quantum wavefunctions in the two superconductors. When these junctions are embedded in a superconducting loop, the resulting system becomes sensitive to external magnetic flux through the phenomenon of quantum interference. This interference causes the supercurrent in the loop to oscillate as a function of the enclosed magnetic flux, enabling the precise detection of changes in magnetic fields.

6.3.2 Types of SQUIDs and Their Operation

There are two main types of SQUIDs: direct current (DC) SQUIDs and radio-frequency (RF) SQUIDs. DC SQUIDs consist of a superconducting loop with two Josephson junctions and operate by measuring the voltage across the junctions as a function of the applied magnetic flux. These devices exhibit a periodic voltage-flux characteristic, allowing them to detect flux changes smaller than a single flux quantum. RF SQUIDs, on the other hand, use a single Josephson junction and are coupled to an RF tank circuit. They are typically simpler in design but slightly less sensitive than DC SQUIDs. In both cases, the devices function by converting magnetic flux into an electrical signal, which can then be amplified and analyzed. The extreme sensitivity of SQUIDs arises from their ability to detect changes in magnetic flux as small as a fraction of a magnetic flux quantum, which is approximately 2.07 × 10^-15 weber.

6.3.3 Applications in Biomagnetism and Medical Diagnostics

One of the most prominent applications of SQUIDs is in the field of biomagnetism, particularly in techniques such as magnetoencephalography (MEG) and magnetocardiography (MCG). MEG measures the tiny magnetic fields produced by neuronal activity in the brain, providing high temporal resolution data that complement other neuroimaging techniques such as MRI and PET. SQUIDs are uniquely capable of detecting these minute signals, which are on the order of femtoteslas, making them indispensable in clinical and research settings. MCG similarly measures the magnetic fields generated by the heart, allowing for noninvasive cardiac diagnostics. The use of SQUIDs in these areas enables early detection of neurological and cardiac disorders and contributes significantly to the understanding of brain function and heart health.

6.3.4 Geophysical and Environmental Applications

SQUIDs have also been applied extensively in geophysical surveys and environmental monitoring. In mineral exploration and archaeology, SQUID magnetometers can detect subtle magnetic anomalies associated with ore bodies or buried artifacts. Their high sensitivity allows geophysicists to create detailed magnetic maps of the subsurface, improving the accuracy of resource estimation and archaeological site investigations. Additionally, SQUIDs can be used to measure the Earth's magnetic field with great precision, contributing to studies in geomagnetism and space weather. Environmental scientists also use SQUID-based instruments to detect trace magnetic particles in the atmosphere or in biological samples, offering insights into pollution levels and their sources.

6.3.5 Fundamental Physics and Quantum Research

In fundamental physics, SQUIDs are invaluable tools for exploring quantum phenomena and testing the boundaries of known physical laws. They are used in experiments involving quantum coherence, quantum entanglement, and the measurement of fundamental constants. For example, SQUIDs play a role in the detection of exotic particles and forces, such as axions or dark matter candidates, by enabling extremely sensitive measurements of induced magnetic fields. In quantum computing, SQUIDs are often used as readout devices for superconducting qubits, which form the basis of some of the most advanced quantum processors currently in development. Their ability to function at cryogenic temperatures and integrate with superconducting circuits makes them ideally suited for these applications.

6.3.6 Advances in SQUID Technology and Miniaturization

Recent advancements in materials science and microfabrication have led to significant improvements in SQUID technology. Modern SQUIDs are often fabricated using thin-film deposition techniques and lithography, enabling miniaturization and integration into compact sensor arrays. These developments have allowed the creation of multi-channel SQUID systems for imaging applications, such as large-scale MEG systems that can simultaneously monitor the activity of many brain regions. Efforts are also underway to develop portable and wearable SQUID systems, which would broaden their use in clinical and field settings. Innovations in cryogenic cooling, including the use of high-temperature superconductors, have also reduced the operational complexity and cost of SQUID systems, making them more accessible for broader applications.

6.3.7 Challenges and Limitations

Despite their remarkable sensitivity, SQUIDs face several practical challenges. The need for cryogenic temperatures, often requiring liquid helium or advanced cryocoolers, adds complexity and cost to their operation. Efforts to use high-temperature superconductors have made progress, but such systems generally exhibit lower sensitivity compared to their low-temperature counterparts. Moreover, SQUIDs are susceptible to magnetic noise from environmental sources, necessitating the use of magnetically shielded environments or sophisticated noise-cancellation techniques. Their fabrication and calibration require high precision and expertise, which can limit their widespread deployment outside specialized laboratories or medical facilities.

6.3.8 Future Directions and Emerging Applications

Looking forward, the future of superconducting sensors and SQUIDs is promising, driven by ongoing research into novel superconducting materials, advanced fabrication methods, and innovative applications. One exciting direction is the use of SQUIDs in quantum sensing networks, where arrays of sensors could provide distributed, high-resolution measurements for a range of scientific and industrial applications. Another emerging area is hybrid quantum systems that integrate SQUIDs with other quantum devices such as optomechanical resonators or spin qubits, potentially enabling new modes of information processing and sensing. The combination of artificial intelligence with SQUID data analysis could also enhance the interpretation of complex datasets in fields like neuroscience and geophysics. As these technologies mature, SQUIDs are expected to play a central role in next-generation instrumentation for both fundamental science and practical applications.

6.4 OPTOMECHANICS IN QUANTUM SENSING

Optomechanics, a field at the intersection of optics and mechanics, has emerged as a powerful platform for exploring quantum phenomena and advancing quantum sensing technologies. It involves the interaction between light and mechanical motion, typically within microscale or nanoscale systems where optical fields can influence mechanical elements such as mirrors, membranes, or cantilevers. In quantum sensing, optomechanical systems offer remarkable sensitivity in detecting extremely small displacements, forces, and changes in environmental parameters by harnessing the quantum properties of both light and mechanical motion. As researchers strive to push the boundaries of measurement precision, optomechanical sensors have shown significant promise in applications ranging from gravitational wave detection to inertial navigation and magnetic field sensing.

Optomechanics has established itself as a cornerstone of quantum sensing, offering unmatched sensitivity and versatility by exploiting the quantum interaction between light and mechanical motion. From detecting gravitational waves to enabling GPS-free navigation and exploring fundamental quantum limits, optomechanical sensors are revolutionizing the precision with which we interact with the physical world. Despite challenges such as thermal noise and decoherence, ongoing innovations in cooling, materials, and integration promise to enhance the capabilities and applicability of these systems. As optomechanics continues to evolve, it is set to play a central role in shaping the future of quantum measurement and sensing technologies.

6.4.1 Fundamentals of Optomechanical Coupling

At the heart of optomechanics lies the principle of radiation pressure, the force exerted by light on a physical object. When light is confined in an optical cavity with a movable mirror or mechanical resonator, its intensity can modulate the mechanical motion of the object. In return, the motion of the mechanical element alters the cavity length, thereby affecting the resonance condition of the light. This mutual coupling gives rise to a dynamic interplay between optical and mechanical degrees of freedom.

In a quantum regime, this interaction becomes a source of entanglement and can be described by quantized optical and phononic modes. The Hamiltonian governing such systems typically includes terms that represent both linear and nonlinear coupling, depending on the specific design and operational regime of the optomechanical system.

6.4.2 Reaching the Quantum Regime in Optomechanics

For optomechanical systems to be useful in quantum sensing, they must operate near or within the quantum regime, where quantum fluctuations dominate the behavior of the system. Achieving this requires cooling the mechanical resonator to its ground state, typically through techniques such as cryogenic cooling or optomechanical sideband cooling. Sideband cooling involves detuning the driving laser relative to the optical cavity's resonance frequency to enable the transfer of energy from the mechanical mode to the optical field, effectively reducing the phonon population of the resonator. When ground-state cooling is accomplished, the system exhibits quantum coherence and low thermal noise, making it highly suitable for detecting weak signals with high precision.

6.4.3 Quantum Measurement and Sensitivity Enhancement

One of the principal applications of optomechanical systems in sensing is in the precise measurement of displacement, force, and acceleration. By monitoring the phase or intensity changes of light in the cavity, one can infer minute shifts in the position of the mechanical element. The sensitivity of such measurements is fundamentally limited by quantum noise, particularly shot noise and back-action noise. However, optomechanical systems can exploit quantum correlations to beat the standard quantum limit, reaching or even surpassing the Heisenberg limit. Techniques such as quantum squeezing and back-action evasion allow the extraction of more information than would otherwise be possible in classical regimes. This capability positions optomechanical sensors as key tools in scenarios demanding ultra-high precision.

6.4.4 Optomechanics in Gravitational Wave Detection

One of the most celebrated uses of optomechanics in quantum sensing is in gravitational wave observatories such as LIGO and Virgo. These facilities employ large-scale interferometers where mirrors suspended at the ends of long arms form optical cavities. As gravitational waves pass through Earth, they induce minute changes in the distances between these mirrors, which are detected through the resulting interference patterns of laser beams. The mirrors themselves act as mechanical resonators, and their motion in response to radiation pressure must be controlled and understood with quantum precision. Optomechanical theories and techniques have been essential in enhancing the sensitivity of these detectors, helping scientists make the first direct observations of gravitational waves and opening a new era of astronomy.

6.4.5 Inertial and Force Sensing Applications

Beyond astrophysics, optomechanical systems are being developed for practical sensing applications in areas such as navigation and force detection. Optomechanical accelerometers and gyroscopes leverage the sensitivity of mechanical resonators to external forces and torques. When coupled with high-finesse optical cavities, these devices can measure accelerations and rotations with extremely low noise and high bandwidth. This capability is particularly valuable in environments where conventional GPS signals are unavailable, such as in submarines or spacecraft. In force sensing, optomechanical devices can detect weak interactions at the nanoscale, including those involved in atomic force microscopy or biomolecular interactions, making them useful in material science and life sciences.

6.4.6 Thermal Noise and Decoherence Challenges

While optomechanical systems hold immense potential, they are not without challenges. One major limitation is thermal noise, which arises from the random motion of the mechanical element due to its interaction with the surrounding environment. This noise can obscure quantum signals and reduce measurement sensitivity. To mitigate this, researchers employ vacuum chambers, cryogenic cooling, and material engineering to isolate the mechanical elements and reduce their coupling to thermal baths. Another challenge is decoherence, where quantum superpositions are destroyed by interactions with the environment. Maintaining quantum coherence over long timescales requires careful design of the system's geometry and materials, as well as precise control over experimental parameters.

6.4.7 Integration with Other Quantum Technologies

A promising avenue in the evolution of optomechanical sensing is its integration with other quantum technologies, including superconducting circuits, atomic ensembles, and spin systems. Hybrid systems can combine the strengths of different platforms, such as the long coherence times of atoms and the strong coupling of photons in optical cavities. For instance, mechanical resonators can be used as intermediaries to transfer quantum states between microwave and optical photons, facilitating quantum communication over different frequency domains. In sensing, such hybrid architectures could enable multifunctional devices that simultaneously measure magnetic fields, electric fields, and temperature with quantum-limited sensitivity.

6.4.8 Miniaturization and On-Chip Integration

The advancement of nanofabrication techniques has enabled the creation of integrated optomechanical devices on chips, bringing the advantages of miniaturization, scalability, and robustness. Photonic crystal cavities, microtoroid resonators, and nanobeams have been engineered to support high-quality optical and mechanical modes, enhancing the optomechanical interaction strength. On-chip optomechanical sensors are attractive for portable and distributed sensing applications, including environmental monitoring, structural health diagnostics, and wearable devices. Their compact size also allows them to be deployed in large arrays for parallel sensing tasks or spatially resolved measurements.

6.4.9 Future Prospects and Emerging Directions

Looking ahead, the future of optomechanics in quantum sensing appears rich with potential. As researchers continue to refine cooling and isolation techniques, optomechanical devices are expected to reach new levels of sensitivity and versatility. One exciting direction is the use of levitated optomechanics, where nanoparticles are trapped and manipulated by optical fields in vacuum, eliminating clamping losses and enabling unprecedented isolation from environmental noise. Another promising area is quantum-enhanced imaging, where optomechanical sensors could be used to detect weak optical signals with spatial resolution beyond the diffraction limit. The ongoing convergence of quantum optics, nanotechnology, and materials science is likely to yield novel sensor architectures and expand the range of measurable quantities.

6.5 COLD ATOM SENSORS

Cold atom sensors represent a significant advancement in precision measurement, utilizing atoms cooled to near absolute zero to detect and measure fundamental physical quantities with unprecedented accuracy. At these ultra-low temperatures, typically achieved using laser cooling and magnetic trapping techniques, atoms move extremely slowly, which allows for their quantum properties to dominate. These quantum characteristics are harnessed in cold atom sensors to achieve exceptional sensitivity in measuring acceleration, rotation, gravity, magnetic fields, and time. This technology has broad applications across physics, navigation, geophysics, and fundamental science. Cold atom sensors stand at the forefront of quantum technology, offering unprecedented precision and stability in the measurement of inertial, gravitational, and temporal quantities. Their unique reliance on the quantum behavior of ultra-cold atoms has opened new pathways in navigation, earth observation, timekeeping, and fundamental research. As engineering and fabrication techniques continue to evolve, cold atom sensors are poised to transition from laboratory curiosities to ubiquitous tools in scientific, industrial, and defense settings. Their development not only exemplifies the practical impact of quantum mechanics but also charts a promising course toward the next generation of sensing technologies.

6.5.1 Principles of Cold Atom Sensing

The operational foundation of cold atom sensors is based on matter-wave interferometry, a concept derived from quantum mechanics which treats atoms as waves rather than particles. In this framework, atoms are split into superpositions of states and allowed to evolve along different paths before being recombined. The resulting interference pattern encodes information about the forces or fields the atoms experienced during their evolution. This approach is analogous to optical interferometry, but with atoms replacing photons. The remarkable advantage of using atoms lies in their intrinsic properties, such as mass and well-known internal transitions, which make them highly sensitive to inertial and gravitational effects.

6.5.2 Techniques for Atom Cooling and Trapping

To enable precise measurements, atoms must be cooled to microkelvin or nanokelvin temperatures. Laser cooling techniques, such as Doppler cooling and sub-Doppler mechanisms, are typically employed in conjunction with magneto-optical traps (MOTs) that confine and cool atoms using counter-propagating laser beams and magnetic field gradients. Once cooled, the atoms are often loaded into optical or magnetic traps for further manipulation.

Evaporative cooling can be applied to reach even lower temperatures, sometimes achieving Bose-Einstein condensation, where a large fraction of atoms occupies the lowest quantum state. This extreme cooling minimizes thermal noise and motion, which is crucial for accurate interferometric sensing.

6.5.3 Atom Interferometry in Inertial Sensing

One of the most prominent applications of cold atom sensors is in inertial navigation. In an atom interferometer designed for accelerometry or gyroscopy, a sequence of laser pulses—often involving Raman or Bragg transitions—is used to split, redirect, and recombine atomic wave packets.

The phase shift in the resulting interference pattern is directly related to acceleration or rotation experienced by the atoms. This technique provides a compact, drift-free alternative to classical gyroscopes and accelerometers, making cold atom sensors ideal for autonomous navigation in submarines, spacecraft, and other GPS-denied environments. Their accuracy and stability outperform traditional systems, especially in long-term operation.

6.5.4 Cold Atom Gravimeters and Gradiometers

Gravimetry is another field where cold atom sensors have made a profound impact. Cold atom gravimeters measure the local gravitational acceleration by observing the phase difference accumulated in an atom interferometer due to the gravitational potential. These instruments have been used in geophysical surveys to map underground structures, monitor volcanic activity, and track groundwater movements. Additionally, cold atom gradiometers, which measure spatial variations in gravity, provide enhanced capabilities for detecting mass distributions over small distances. These devices are critical in applications requiring high-resolution gravity mapping, such as mineral exploration and environmental monitoring.

6.5.5 Advancements in Atomic Clocks

Cold atoms have revolutionized the field of timekeeping through their application in atomic clocks. The precision of an atomic clock is governed by the sharpness of the atomic transition frequency and the duration over which the transition is observed. Cold atoms, due to their slow motion, allow for longer interrogation times and narrower transition linewidths, significantly improving timekeeping accuracy. Cesium fountain clocks and optical lattice clocks using strontium or ytterbium atoms have achieved extraordinary precision, becoming the backbone of international time standards. These clocks are essential not only for defining the second but also for enabling technologies such as satellite navigation and telecommunications.

6.5.6 Quantum Noise Reduction and Coherence Control

The sensitivity of cold atom sensors is ultimately constrained by quantum noise, including projection noise and quantum decoherence. To overcome these limitations, researchers employ techniques like spin squeezing and entanglement, which allow measurement precision to exceed the standard quantum limit. Maintaining coherence of atomic states during measurement is also crucial. This is achieved by minimizing interactions with the environment and using robust laser systems and vacuum conditions. Advances in coherence control have allowed cold atom sensors to operate over extended durations and achieve higher stability, especially in mobile or field-deployable platforms.

6.5.7 Portable and Field-Deployable Cold Atom Sensors

Traditionally, cold atom experiments required large, complex laboratory setups. However, recent technological progress has led to the development of compact, transportable cold atom sensors. These devices integrate miniaturized vacuum systems, fiber-coupled lasers, and microfabricated components to reduce size and power requirements. Such innovations have enabled cold atom sensors to be deployed in diverse environments, from onboard aircraft and ships to remote field stations. Their ruggedness and autonomy make them suitable for real-time environmental monitoring, planetary exploration, and defense applications where conventional sensors may fail due to harsh conditions or lack of calibration standards.

6.5.8 Cold Atom Sensors in Fundamental Physics

Beyond practical sensing applications, cold atom sensors serve as powerful tools in probing fundamental physics. They have been used to test the equivalence principle, a cornerstone of general relativity, by comparing the free-fall acceleration of different atomic species. Cold atom experiments also offer platforms for detecting tiny variations in fundamental constants and exploring possible extensions to the

Standard Model, such as searching for dark matter or new gravitational interactions. Their sensitivity to minute perturbations makes them ideal for detecting effects that are otherwise inaccessible to classical instruments.

6.5.9 Challenges and Future Directions

Despite their numerous advantages, cold atom sensors face challenges related to system complexity, operational stability, and sensitivity to environmental disturbances. The need for ultra-high vacuum, stable laser systems, and precise control over magnetic and electric fields makes them technically demanding. Ongoing research aims to simplify sensor architecture, improve robustness, and enhance integration through atom chips and photonic circuits. Future directions include the deployment of cold atom sensors in space missions, where their isolation from terrestrial noise sources could unlock new insights into Earth science and fundamental physics. In particular, missions like STE-QUEST and CAL (Cold Atom Laboratory) aboard the International Space Station are pioneering this frontier.

6.6 SPIN-BASED QUANTUM SENSORS

Spin-based quantum sensors represent a critical frontier in quantum technology, exploiting the intrinsic angular momentum of particles, known as spin, to detect and measure physical quantities with exceptional precision. These sensors use the quantum mechanical property of spin in particles such as electrons, nuclei, or atoms to probe magnetic fields, electric fields, temperature, and more. Their operation hinges on the high sensitivity of spin states to environmental changes, and the ability to control and read out these states with remarkable accuracy. As a result, spin-based quantum sensors have emerged as versatile tools in fundamental physics, material science, biomedical imaging, and beyond. Spin-based quantum sensors exemplify the transformative potential of quantum technologies in precision measurement. By exploiting the quantum coherence and sensitivity of spin systems, these sensors achieve exceptional resolution in detecting magnetic fields, electric fields, temperature, and other physical quantities. Their versatility and miniaturization make them ideal for diverse applications, from fundamental physics experiments to practical uses in medicine, materials science, and navigation. As the technology matures and overcomes current limitations, spin-based quantum sensors are poised to become indispensable tools across scientific and industrial domains, marking a profound shift in how we observe and interact with the world.

6.6.1 The Quantum Nature of Spin

At the heart of spin-based quantum sensors lies the quantum property of spin, which is not a classical spinning motion but rather a fundamental quantum number characterizing particles. Spin exists in discrete states, such as "up" or "down" for spin-½ particles, and these states can be manipulated and superposed in quantum systems. In quantum sensing, spin states are coherently prepared, evolve in response to external perturbations, and are then measured to extract information about those perturbations. The coherence of spin states and the sharp energy differences between them make them ideal for precise measurement applications.

6.6.2 Detection Mechanisms in Spin-Based Sensing

The operation of a spin-based quantum sensor typically involves initializing the spin state, allowing it to evolve under the influence of an external field, and performing a readout to determine the resulting state. Magnetic resonance techniques, including electron spin resonance (ESR) and nuclear magnetic resonance (NMR), are foundational to this process.

A small perturbation in a magnetic or electric field alters the energy levels of the spin system, which in turn shifts the resonance condition. These shifts are detected with high precision, enabling the sensor to infer minute changes in its environment. The resolution of this detection is enhanced by quantum coherence and long spin relaxation times, which help preserve the spin information during measurement.

6.6.3 Materials and Platforms for Spin-Based Sensors

Several materials systems are employed for spin-based quantum sensing, each offering specific advantages. One prominent platform is the nitrogen-vacancy (NV) center in diamond, where a nitrogen atom and a neighboring vacancy replace two carbon atoms in the diamond lattice. This defect forms a stable electronic spin system that can be optically initialized and read out even at room temperature, making NV centers highly suitable for biological and material applications. Other platforms include phosphorus-doped silicon, rare-earth ions in crystals, and trapped ions or atoms. These materials provide different trade-offs in terms of coherence time, operational temperature, and integration into devices.

6.6.4 Applications in Magnetometry

Spin-based quantum sensors are especially powerful in magnetometry, the measurement of magnetic fields. Using the Zeeman effect, in which energy levels split in the presence of a magnetic field, spin systems can detect even picotesla-level fields. NV center magnetometers, for example, are employed to image neuronal activity in living tissues, map magnetic domains in ferromagnetic materials, and detect nanoscale magnetic particles. Unlike traditional magnetometers, these sensors offer high spatial resolution and operate effectively under ambient conditions. This opens new avenues for non-invasive diagnostics, geophysical exploration, and quality control in manufacturing processes.

6.6.5 Electric Field and Temperature Sensing

Though magnetic fields are the primary focus of many spin-based sensors, they can also be tuned to respond to electric fields and temperature changes. Stark shifts in energy levels due to electric fields, as well as thermally induced fluctuations in spin resonance, allow the same spin systems to serve as versatile detectors. NV centers, again, play a central role in these domains, offering nanoscale resolution in temperature mapping, particularly within biological samples. This capability enables real-time monitoring of cellular metabolism, thermal diffusion, and even intracellular reactions, demonstrating the interdisciplinary utility of spin-based sensing.

6.6.6 Biological and Medical Applications

The use of spin-based quantum sensors in biological contexts is a rapidly growing field, driven by the non-invasive and high-resolution nature of these devices. NV center sensors can be embedded into nanodiamonds and introduced into cells or tissues, where they monitor magnetic signals or temperature variations. These sensors can track ion concentrations, membrane potentials, and protein dynamics, offering unprecedented insight into cellular function and disease mechanisms. In neuroscience, they are used to investigate magnetic signals from neurons without the need for electrodes or invasive probes. As sensor miniaturization advances, their role in real-time diagnostics and drug development is expected to expand significantly.

6.6.7 Spin-Based Sensors in Materials Science

In materials research, spin-based sensors are deployed to study the magnetic, electronic, and thermal properties of solids at microscopic scales. They enable nanoscale imaging of magnetic vortices, domain structures, and current flow in advanced materials such as superconductors, topological insulators, and spintronic devices. By detecting local magnetic noise or spin fluctuations, researchers gain insight into material behavior that would be inaccessible through macroscopic measurements. These capabilities are crucial for designing new materials with tailored quantum properties and for optimizing the performance of quantum computers and memory devices.

6.6.8 Quantum Enhancement and Entanglement

To surpass the limits imposed by classical physics, spin-based sensors can be enhanced using quantum entanglement and squeezing techniques. Entangled spin states allow for collective measurements that reduce noise and increase sensitivity beyond the standard quantum limit. These improvements are vital in applications requiring ultra-precise measurements, such as tests of fundamental physics or detection of weak gravitational or magnetic fields. Quantum error correction and feedback mechanisms are also being explored to extend the operational coherence time of spin sensors, making them even more robust against environmental noise and decoherence.

6.6.9 Challenges in Implementation

Despite their promise, spin-based quantum sensors face technical and practical challenges. Maintaining spin coherence in noisy or fluctuating environments requires advanced shielding, cooling, or dynamic decoupling techniques. Sensor calibration, especially at the nanoscale, can be complex due to inhomogeneities in the sensing environment. Furthermore, integrating spin sensors into portable or wearable devices involves overcoming hurdles related to size, power consumption, and signal processing. Nevertheless, ongoing research is steadily addressing these challenges through improved materials, chip-scale integration, and machine learning-based control systems.

6.6.10 Future Outlook and Innovations

The future of spin-based quantum sensing lies in expanding the range of measurable quantities, improving device integration, and scaling up for networked or distributed sensing applications. Developments in hybrid quantum systems, where spin sensors are coupled to photonic or mechanical resonators, promise enhanced sensitivity and broader operational regimes. In addition, quantum networking may enable remote spin sensors to function collectively, offering new modalities for environmental monitoring or global positioning systems. The integration of spin-based sensors with classical electronics and emerging quantum processors will also pave the way for intelligent sensing systems capable of real-time decision-making and adaptation.

6.7 PHOTONIC SENSORS AND QUANTUM LIGHT SOURCES

Photonic sensors and quantum light sources are pivotal components in the rapidly advancing field of quantum technology. These technologies leverage the unique properties of light and photons to achieve unparalleled sensitivity, precision, and control in a variety of measurement applications. Photonic sensors exploit the behavior of light in both classical and quantum regimes, often harnessing quantum mechanical phenomena such as superposition, entanglement, and interference to achieve performance that far exceeds that of traditional sensors. Quantum light sources, on the other hand, generate non-classical states of light, such as squeezed states or entangled photon pairs, which can be used to enhance the sensitivity of measurements and enable new types of quantum-based sensing techniques. Together, these technologies form the backbone of many emerging quantum applications in fields ranging from communication and navigation to medicine and environmental monitoring. Photonic sensors and quantum light sources are at the forefront of the quantum revolution, offering unprecedented sensitivity and precision for a wide range of applications. By harnessing the unique properties of light, such as superposition, entanglement, and squeezing, these technologies enable measurements that go beyond the capabilities of classical systems. From quantum metrology and imaging to secure communication and environmental monitoring, photonic sensors and quantum light sources are poised to drive innovations across many fields. As research progresses, overcoming the current challenges in photon generation, integration, and scaling will unlock even more transformative applications, ushering in a new era of quantum-enhanced technologies.

6.7.1 The Role of Photonic Sensors in Quantum Sensing

Photonic sensors are devices that detect and measure light, often with extreme sensitivity to small changes in intensity, wavelength, polarization, or phase. In the context of quantum sensing, these sensors exploit the fundamental properties of light and photons to detect phenomena with much higher precision than classical systems. One of the key advantages of photonic sensors is their ability to work in the quantum regime, where the superposition of quantum states and interference effects enable measurements beyond the capabilities of classical systems. These sensors are integral to a wide range of applications, including quantum imaging, spectrometry, magnetometry, and gravitational wave detection. Photonic sensors can be based on a variety of technologies, including avalanche photodiodes (APDs), superconducting nanowires, and photomultiplier tubes (PMTs). These sensors are often optimized for specific quantum properties of light, such as photon counting, which allows for extremely accurate measurements of light intensity at the single-photon level. The ability to detect single photons with high fidelity is crucial in many quantum technologies, including quantum key distribution for secure communications and quantum metrology for precise measurement of physical quantities like time, distance, and magnetic fields.

6.7.2 Quantum Light Sources: Fundamentals and Types

Quantum light sources are devices that generate non-classical states of light, which are essential for many quantum sensing applications. Unlike classical light sources, which emit coherent or incoherent light with well-defined properties, quantum light sources produce light in states that are not described by classical physics. These non-classical states of light have unique properties that can enhance the performance of photonic sensors and enable new types of measurements. One of the most common types of quantum light sources is the squeezed light source, which produces light with reduced uncertainty in one property (such as phase) at the expense of increased uncertainty in another (such as amplitude). Squeezed light can be used to improve the sensitivity of photonic sensors, particularly in measurements that are limited by shot noise or thermal fluctuations. By reducing the noise in one quadrature of the light field, squeezed states allow for more precise measurements of weak signals, such as tiny magnetic fields or gravitational waves. Another important class of quantum light sources is entangled photon pairs, which are generated through processes such as spontaneous parametric down-conversion (SPDC). Entangled photons exhibit strong correlations in their properties, such as polarization or phase, even when they are separated by large distances. These entangled states can be used to enhance the performance of photonic sensors by allowing for quantum interference effects that improve the precision of measurements. Entangled photon sources are used in applications such as quantum imaging, quantum-enhanced interferometry, and quantum metrology.

6.7.3 Applications of Photonic Sensors and Quantum Light Sources

The combination of photonic sensors and quantum light sources has led to groundbreaking advancements in a variety of scientific and technological fields. One of the most promising applications is in quantum metrology, where the enhanced precision of quantum measurements allows for the determination of physical constants and quantities with unprecedented accuracy. For example, quantum light sources are used in highly sensitive interferometers for detecting gravitational waves, as demonstrated by the Laser Interferometer Gravitational-Wave Observatory (LIGO). By utilizing squeezed light in these interferometers, researchers have been able to reduce the noise that limits the sensitivity of classical detectors, enabling the detection of minuscule distortions in spacetime caused by distant astrophysical events.

Quantum-enhanced imaging and spectroscopy are other areas where photonic sensors and quantum light sources are making a significant impact. In biological and medical imaging, quantum sensors are being developed to enable high-resolution, non-invasive imaging of living tissues at the single-photon level. These sensors are capable of detecting weak fluorescence signals, which can provide valuable information about cellular processes, protein interactions, and other biological phenomena. Quantum-enhanced imaging techniques are also being explored for use in environmental monitoring, where they could help detect pollutants, monitor atmospheric conditions, or map the Earth's surface with unprecedented accuracy. Quantum light sources are also being used in the field of quantum communication, where they form the basis for secure communication protocols such as quantum key distribution (QKD). In QKD, entangled photon pairs are used to establish secure communication channels that are immune to eavesdropping.

The use of quantum light sources ensures that any attempt at intercepting the communication would disturb the quantum state of the photons, making eavesdropping detectable. This technology is poised to revolutionize the field of secure communications, providing a level of security that is fundamentally unattainable with classical systems.

6.7.4 Challenges and Limitations

Despite their immense potential, photonic sensors and quantum light sources face several challenges that must be addressed to fully realize their capabilities. One of the main challenges is the generation of high-quality quantum light sources that can operate efficiently at scale. For example, while squeezed light and entangled photon pairs have been demonstrated in laboratory settings, their generation is often complex and requires precise control of the physical processes involved. Scaling up these technologies to produce large numbers of photons with high fidelity is an ongoing area of research. Another challenge is the integration of quantum light sources with photonic sensors in practical devices. Many quantum sensing applications require the integration of both the light source and the sensor into a single compact system. This integration is challenging because quantum light sources often require specialized conditions, such as low temperatures or vacuum environments, while photonic sensors may operate optimally at room temperature. Developing hybrid systems that can operate under a wide range of conditions while maintaining quantum coherence is a key focus of current research.

6.7.5 Future Directions

The future of photonic sensors and quantum light sources is bright, with numerous exciting possibilities on the horizon. One promising direction is the development of quantum-enhanced sensing networks, where multiple sensors are linked together to create a distributed sensing system. These networks could be used for a variety of applications, including environmental monitoring, geophysical exploration, and healthcare. The integration of quantum light sources with photonic sensors in these networks could enable real-time, high-resolution measurements over large areas, with applications ranging from climate change monitoring to remote sensing of natural resources. Advances in photonic materials and integrated photonics are also expected to drive the development of more efficient and compact quantum sensors. By leveraging advancements in nanophotonics and photonic chip fabrication, researchers are working to create miniature quantum sensors that can be easily integrated into everyday devices. These sensors could have applications in consumer electronics, wearable health monitors, and industrial sensors, making quantum sensing technologies more accessible to a wider audience.

CHAPTER 7

Quantum Sensor Performance and Sensitivity

7.1 QUANTUM SENSOR SENSITIVITY AND PRECISION

Quantum sensors represent a breakthrough in measurement technology, leveraging the principles of quantum mechanics to achieve unprecedented levels of sensitivity and precision. Unlike classical sensors, which are constrained by the limits of classical physics, quantum sensors exploit quantum phenomena such as superposition, entanglement, and interference to provide highly accurate and detailed measurements. These devices have found applications across various fields, including precision measurement, navigation, imaging, and communication. The sensitivity and precision of quantum sensors are critical to their effectiveness, and understanding the underlying mechanisms that govern these properties is essential for optimizing their performance in real-world applications. Quantum sensor sensitivity and precision represent two of the most significant advantages of quantum sensing technologies. By exploiting the principles of quantum mechanics, such as superposition, entanglement, and coherence, quantum sensors achieve unprecedented levels of measurement accuracy that surpass the capabilities of classical sensors. As quantum sensing technologies continue to evolve, their ability to measure extremely small changes in physical quantities will enable breakthroughs in fields ranging from fundamental physics and medical diagnostics to environmental monitoring and navigation. Understanding and optimizing the sensitivity and precision of quantum sensors will be crucial in unlocking their full potential and realizing the many benefits they offer for both scientific research and practical applications.

7.1.1 The Concept of Sensitivity in Quantum Sensing

Sensitivity refers to the ability of a sensor to detect small changes in the quantity it is measuring. In the context of quantum sensors, sensitivity often refers to the smallest detectable signal or the smallest change in a physical quantity, such as a magnetic field, temperature, or gravitational wave, that the sensor can reliably measure. Quantum sensors are uniquely positioned to achieve extremely high sensitivity due to their ability to exploit quantum mechanical effects. For example, in quantum metrology, sensors can detect minute variations in physical quantities that would be undetectable by classical sensors. One of the key principles that enhances the sensitivity of quantum sensors is quantum superposition. By placing a quantum system in a superposition of states, sensors can effectively amplify the measurable effects of a small perturbation. This principle is often employed in quantum interferometers, where the interference of quantum states can be used to measure very small changes in phase, position, or frequency. Additionally, the use of squeezed states of light, which reduce the uncertainty in one property (such as phase) while increasing uncertainty in another (such as amplitude), can further enhance the sensitivity of quantum sensors.

7.1.2 Precision in Quantum Sensing

While sensitivity refers to the smallest detectable signal, precision is a measure of how consistently a sensor can produce the same measurement under repeated observations. Precision is vital for ensuring that measurements are accurate and reliable over time. In quantum sensing, precision is closely linked to the ability of the sensor to minimize errors, fluctuations, and noise that may affect the measurement outcome. Quantum sensors are capable of achieving extremely high precision because they can harness the properties of quantum systems that are not available to classical sensors.

The precision of a quantum sensor is often constrained by fundamental limits, such as the Heisenberg uncertainty principle, which defines the relationship between the uncertainty in the measurement of conjugate variables, such as position and momentum or energy and time. The Heisenberg uncertainty principle sets a fundamental lower bound on the precision with which certain pairs of measurements can be made. However, quantum sensors can exploit strategies such as squeezing and entanglement to reduce uncertainties in specific measurement properties, thereby improving precision beyond the capabilities of classical sensors.

7.1.3 The Role of Quantum Entanglement in Precision

Quantum entanglement plays a central role in enhancing both the sensitivity and precision of quantum sensors. Entangled states are a special type of quantum state in which the properties of two or more particles become correlated in such a way that the state of one particle cannot be described independently of the state of the other. This non-local correlation between particles can be used to improve the precision of measurements by enabling interference effects that amplify small changes in the measured quantity. In quantum interferometry, entangled particles are used to create interference patterns that are highly sensitive to small changes in the phase or position of the particles. This sensitivity can be leveraged to achieve measurements with extreme precision, surpassing the limitations imposed by classical interferometers. Entanglement also enables the use of quantum metrology techniques, which allow for the measurement of fundamental physical constants with greater precision than classical methods.

7.1.4 Quantum Coherence and Its Impact on Sensitivity and Precision

Quantum coherence is another crucial factor that impacts the sensitivity and precision of quantum sensors. Coherence refers to the ability of a quantum system to maintain a well-defined phase relationship between its components over time. In quantum sensing, coherence is essential for ensuring that the quantum states used in measurements remain stable and correlated, which enables high precision and sensitivity. Loss of coherence, known as decoherence, can significantly degrade the performance of a quantum sensor. Decoherence can occur due to interactions with the environment, such as thermal fluctuations, electromagnetic radiation, or other sources of noise. To maintain high sensitivity and precision, quantum sensors must be carefully engineered to minimize decoherence and preserve coherence over long periods of time. Techniques such as dynamical decoupling, which involves applying sequences of pulses to protect the quantum state from environmental disturbances, are often used to mitigate the effects of decoherence in quantum sensors.

7.1.5 Quantum Noise and Its Influence on Sensor Performance

Quantum noise, also known as shot noise or photon noise, arises from the discrete nature of light and particles at the quantum level. This noise can have a significant impact on the performance of quantum sensors, as it limits both the sensitivity and precision of measurements. Quantum noise is typically characterized by fluctuations in the measurement outcome that arise due to the probabilistic nature of quantum systems. These fluctuations are often most noticeable in systems that rely on photon detection or other quantum measurements. Quantum noise can be reduced using techniques such as squeezed states, which reduce the uncertainty in one quadrature of the quantum state (such as phase) while increasing the uncertainty in the conjugate quadrature (such as amplitude). By squeezing the state of the light or particles used in the sensor, it is possible to reduce the shot noise in the measurement process, thereby enhancing the sensitivity and precision of the sensor. Squeezing is especially useful in applications such as quantum interferometry, where reducing quantum noise can lead to significant improvements in the precision of measurements, such as detecting gravitational waves or measuring small changes in magnetic fields.

7.1.6 Heisenberg and Standard Quantum Limits

The sensitivity and precision of quantum sensors are often compared against two important theoretical limits: the Heisenberg limit and the standard quantum limit (SQL). The standard quantum limit is a measure of the best sensitivity achievable using classical techniques or weak quantum measurements, while the Heisenberg limit represents the ultimate bound on sensitivity imposed by quantum mechanics. At the standard quantum limit, quantum sensors are typically limited by shot noise, which arises from the discrete nature of photons or other quantum particles. The SQL sets the threshold beyond which further improvements in sensitivity require the use of advanced quantum techniques, such as squeezing or entanglement. In contrast, the Heisenberg limit represents the maximum achievable precision, which can be attained through the use of highly entangled quantum states or other advanced quantum techniques. The Heisenberg limit often provides an advantage over classical systems in terms of both sensitivity and precision, allowing quantum sensors to measure quantities with far greater accuracy.

7.1.7 Applications of Quantum Sensor Sensitivity and Precision

The extraordinary sensitivity and precision of quantum sensors have led to their use in a wide range of applications across various scientific and industrial fields. In the realm of fundamental physics, quantum sensors are used to measure physical constants with exceptional precision, testing theories of gravity and quantum mechanics. For example, quantum gravimeters and accelerometers are used to measure gravitational forces with unprecedented accuracy, providing valuable insights into the nature of gravity and enabling the detection of small variations in the Earth's gravitational field. In the field of medicine, quantum sensors are being developed to enhance medical imaging techniques, allowing for higher resolution images and better detection of small changes in biological systems.

Quantum-enhanced magnetic resonance imaging (MRI) and spectroscopy techniques are already being explored for use in non-invasive diagnostics and monitoring. Furthermore, quantum sensors are poised to revolutionize navigation and positioning systems. Quantum accelerometers and gyroscopes are being developed for use in GPS-free navigation systems, offering highly accurate measurements of acceleration and rotation without the need for external references.

7.2 SCALING QUANTUM SENSORS FOR PRACTICAL USE

Quantum sensors have emerged as revolutionary tools for high-precision measurement across various scientific and technological fields. These sensors, based on the principles of quantum mechanics, leverage phenomena like superposition, entanglement, and quantum interference to achieve sensitivity and precision far beyond the capabilities of classical systems. However, despite their remarkable potential, the widespread adoption of quantum sensors for practical, real-world applications faces several challenges. One of the most significant obstacles is the scaling of quantum sensors from laboratory-based prototypes to devices that can be deployed in commercial and industrial environments. This process involves addressing various technical, engineering, and environmental challenges to ensure that quantum sensors can perform effectively outside the controlled conditions of a laboratory. Scaling quantum sensors for practical use is a complex and multifaceted challenge that involves overcoming issues related to size, stability, cost, and environmental sensitivity. However, significant progress is being made in addressing these challenges. Through innovations in miniaturization, error correction, materials science, and manufacturing, quantum sensors are becoming more viable for real-world applications. As these sensors continue to evolve and mature, they hold the potential to revolutionize industries ranging from healthcare and navigation to environmental monitoring and industrial manufacturing. The successful scaling of quantum sensors will open up new frontiers in precision measurement and enable groundbreaking advancements in a wide array of scientific and technological fields.

7.2.1 Challenges in Scaling Quantum Sensors

The first major challenge in scaling quantum sensors is the issue of size. Quantum sensors often require delicate, highly controlled environments to function optimally. For instance, quantum systems like cold atoms or trapped ions typically require ultra-low temperatures, vacuum chambers, and highly stable magnetic or electric fields to maintain coherence and prevent decoherence. These stringent environmental requirements make it difficult to scale quantum sensors into portable or compact devices that can be used in real-world conditions. The size of the apparatus needed to maintain these conditions often limits the practical applications of quantum sensors, especially in fields like portable sensing or mobile health diagnostics. Another challenge in scaling quantum sensors is the sensitivity to environmental noise.

Quantum systems are inherently fragile and can be easily disturbed by external factors such as vibrations, temperature fluctuations, electromagnetic fields, and even cosmic radiation. These external disturbances can cause decoherence, reducing the accuracy and reliability of the measurements. For example, quantum interferometers, which rely on maintaining phase coherence between quantum states, can suffer from environmental noise, making them less effective in non-laboratory settings. Achieving the robustness needed for quantum sensors to function effectively in noisy, uncontrolled environments is a key hurdle to their scalability.

7.2.2 Miniaturization of Quantum Sensors

The miniaturization of quantum sensors is essential to their scalability for practical applications. To move beyond laboratory setups, researchers are developing techniques to make quantum sensors more compact without compromising their performance. One approach is the use of integrated photonics, where quantum states are manipulated and measured on a chip-scale device. Integrated quantum photonic circuits can significantly reduce the size and cost of quantum sensors, making them more suitable for widespread use. Another promising technique for miniaturizing quantum sensors involves using semiconductor-based systems, such as quantum dots or nitrogen-vacancy (NV) centers in diamond. These solid-state systems can be integrated into small, compact devices that are easier to deploy in various environments. NV centers, for instance, are already being used for quantum sensing applications like magnetometry and thermometry, and they offer a path toward miniaturization due to their ability to operate at room temperature without the need for extreme cooling systems. Advances in semiconductor fabrication techniques are expected to further accelerate the miniaturization of quantum sensors, making them more accessible for commercial and industrial use.

7.2.3 Improving Quantum Sensor Stability and Robustness

Stability and robustness are critical factors when scaling quantum sensors for real-world applications. To ensure that quantum sensors can function effectively outside the laboratory, they must be engineered to withstand the rigors of the real world, including temperature fluctuations, vibrations, and electromagnetic interference. This requires developing advanced materials and shielding techniques that can protect the quantum system from environmental disturbances. One promising approach to improving the stability of quantum sensors is the use of quantum error correction methods. These techniques aim to detect and correct errors caused by decoherence or other environmental influences, allowing quantum sensors to maintain high accuracy over time. Quantum error correction is a rapidly developing field, and while it is still in its infancy, it holds the potential to significantly enhance the stability and reliability of quantum sensors in practical settings. Additionally, advances in the design of quantum sensors themselves can contribute to improved robustness. For example, researchers are exploring ways to make quantum sensors less sensitive to environmental noise by designing systems that are inherently less prone to decoherence. By focusing on creating quantum systems that can better tolerate external disturbances, it may be possible to improve the performance of quantum sensors in a wider range of environments.

7.2.4 Cost Reduction and Scalability

The cost of quantum sensors is another significant barrier to their widespread use. Many quantum sensors, especially those that require ultra-high vacuum environments or extreme cooling systems, are expensive to produce and maintain. To make quantum sensors viable for commercial and industrial applications, it is crucial to reduce their cost while maintaining their performance. One way to reduce costs is by developing simpler and more cost-effective quantum systems. For example, as mentioned earlier, NV centers in diamond are a promising candidate for scaling quantum sensors because they can operate at room temperature without requiring expensive cooling equipment. Additionally, advances in semiconductor technology may allow for the creation of quantum sensors that can be mass-produced using existing manufacturing techniques, reducing the cost per unit. Another approach to reducing costs is to improve the manufacturing processes for quantum sensors. Advances in techniques like nanofabrication and photonic integration could help lower the cost of producing quantum sensors by enabling large-scale production of high-quality components. As quantum technologies continue to mature, economies of scale should further reduce costs, making quantum sensors more accessible for a wide range of applications.

7.2.5 Quantum Sensor Networks

To further enhance the utility of quantum sensors, researchers are exploring the idea of creating quantum sensor networks. These networks would consist of multiple quantum sensors working together to provide a more comprehensive and accurate picture of the environment. For instance, a network of quantum gravimeters could be used to monitor subtle changes in the Earth's gravitational field, providing valuable data for geophysical surveys or disaster prediction. Similarly, a network of quantum magnetometers could be used to monitor magnetic fields in a large area for applications such as mineral exploration or environmental monitoring. Quantum sensor networks could offer several advantages over individual quantum sensors, including improved accuracy, redundancy, and resilience to noise. By combining the measurements of multiple sensors, it may be possible to reduce errors and enhance the overall precision of the system. Furthermore, quantum sensor networks could enable new applications, such as quantum-enhanced communication and secure data transmission, by taking advantage of the entanglement and superposition properties of quantum states.

7.2.6 Applications of Scaled Quantum Sensors

As quantum sensors become more scalable, they will find a wide range of practical applications. In the field of navigation, quantum accelerometers and gyroscopes can provide highly accurate measurements of position and orientation, enabling GPS-free navigation in environments where traditional satellite-based systems are unavailable. In healthcare, miniaturized quantum sensors could be used for non-invasive diagnostics, offering new ways to monitor biomolecular processes or detect diseases at an early stage. Other potential applications for scaled quantum sensors include precision measurements in fundamental physics, environmental monitoring, and industrial quality control. For instance, quantum magnetometers could be used to monitor the magnetic fields produced by electrical machinery, helping to identify potential issues before they cause system failures. Quantum thermometers could be used to measure temperature with extreme precision in environments where classical thermometers struggle to perform, such as at the nanoscale or in extreme conditions like deep space.

7.3 STRATEGIES TO OVERCOME DECOHERENCE AND NOISE

Quantum systems are remarkably sensitive to external disturbances, making them highly prone to decoherence and noise. These phenomena can degrade the performance of quantum devices, particularly quantum sensors and quantum computers, which rely on delicate quantum states such as superposition and entanglement. Decoherence occurs when a quantum system interacts with its environment, causing the quantum state to lose its coherence and behave more classically. Noise, on the other hand, refers to random fluctuations in the environment that introduce errors or uncertainties in the measurement process. Both decoherence and noise pose significant challenges for the practical realization of quantum technologies.

Therefore, developing strategies to overcome these obstacles is crucial to advancing quantum devices for real-world applications. Decoherence and noise are among the most significant obstacles to the practical use of quantum systems, particularly in quantum sensors and computers. However, a wide range of strategies are being explored to mitigate these challenges and improve the performance of quantum devices. Techniques such as quantum error correction, topological quantum computing, environmental shielding, quantum feedback, and advances in materials and fabrication all contribute to overcoming the effects of decoherence and noise. As these strategies continue to evolve, it is likely that quantum technologies will become more robust and reliable, paving the way for their widespread adoption in practical applications across a variety of fields.

7.3.1 Understanding Decoherence and Noise

Before delving into strategies for overcoming decoherence and noise, it is important to understand the nature of these phenomena. Decoherence arises due to the entanglement between a quantum system and its environment. When this entanglement occurs, the quantum system's pure state becomes a mixed state, which means the system no longer exhibits purely quantum behavior. In the context of quantum sensors, decoherence can lead to loss of sensitivity and accuracy, as the measurement is no longer based on the precise superposition of quantum states. Noise, on the other hand, refers to the unwanted interactions between a quantum system and the surrounding environment, which can disrupt the measurement process. These environmental factors can be electromagnetic, thermal, or even mechanical in nature. Noise can manifest as random fluctuations that lead to errors in the quantum state, and it can be a source of uncertainty in the output of quantum sensors or computations. Both decoherence and noise can severely limit the practical application of quantum devices, making it necessary to explore effective strategies to mitigate their impact.

7.3.2 Decoherence-Free Subspaces and Error Correction

One of the most promising strategies to overcome decoherence is the concept of decoherence-free subspaces (DFS). This approach involves encoding the quantum information in a way that protects it from specific types of environmental noise. DFS exploits symmetries in the interaction between the quantum system and its environment, effectively shielding the quantum states from the harmful effects of decoherence. By carefully selecting quantum states that are immune to certain types of noise, it becomes possible to maintain coherence even in noisy environments. Another complementary approach to mitigating decoherence is quantum error correction (QEC). In classical computing, error correction involves detecting and correcting errors in bits to ensure reliable operation. Similarly, quantum error correction involves using redundant quantum information to detect and correct errors caused by decoherence or noise. QEC schemes, such as the Shor code or the surface code, can protect quantum information from decoherence by encoding it across multiple physical qubits. Although quantum error correction is resource-intensive and requires significant overhead, it is a powerful tool for maintaining the integrity of quantum information in noisy environments.

7.3.3 Topological Quantum Computing

Topological quantum computing is an innovative strategy that aims to reduce the impact of decoherence by exploiting the topological properties of quantum states. Unlike traditional qubits, which can be sensitive to local perturbations, topological qubits are encoded in global properties of the system, making them more robust against local noise and decoherence. These qubits are based on exotic particles known as anyons, which exhibit non-abelian statistics and can be manipulated through braiding operations. Since the quantum information is stored in the global properties of the system rather than in the individual qubits, topological quantum computing is expected to be less prone to errors caused by decoherence and noise. In practical terms, topological quantum computing could revolutionize the field by providing a more stable platform for quantum computations. Although topological qubits are still in the experimental phase, they hold great potential for overcoming one of the major hurdles in the development of scalable quantum computers and sensors.

7.3.4 Environmental Shielding and Noise Mitigation

While quantum error correction and decoherence-free subspaces focus on mitigating the effects of decoherence, other strategies focus on reducing the noise that causes it in the first place. Environmental shielding is a critical technique in protecting quantum systems from external disturbances such as electromagnetic fields, vibrations, and temperature fluctuations. For instance, quantum sensors, such as those based on trapped ions or superconducting qubits, are often placed inside vacuum chambers to minimize interactions with air molecules. Additionally, these devices may be surrounded by electromagnetic shielding to protect them from external electric and magnetic fields that could introduce noise.

Thermal noise is another common source of disturbance in quantum systems. Low temperatures are essential for reducing thermal fluctuations that can affect quantum coherence. Cryogenic techniques, such as using dilution refrigerators, are often employed to cool quantum systems to near absolute zero temperatures, significantly reducing thermal noise. By isolating quantum devices from external temperature variations and using advanced cooling techniques, the impact of thermal noise can be minimized, allowing quantum states to remain stable for longer periods.

7.3.5 Quantum Sensing with Quantum Feedback

Quantum feedback is a powerful technique for counteracting decoherence and noise by continuously monitoring and correcting the quantum state of the system. In a quantum feedback loop, the system's state is measured in real-time, and feedback is applied to correct any errors or deviations from the desired state. This can be achieved through adaptive measurement strategies, where the quantum system is continually adjusted based on the feedback obtained from the measurements. Quantum feedback has been successfully demonstrated in several quantum systems, such as quantum optomechanics, where feedback is used to stabilize the position of a nanoparticle and mitigate the effects of decoherence. By using real-time monitoring and correction, quantum feedback allows for more precise control over quantum systems, enabling them to maintain coherence for longer durations even in the presence of noise.

7.3.6 Quantum Sensing in Superconducting Qubits

Superconducting qubits, which are widely used in quantum computing and sensing, are particularly susceptible to noise due to their interaction with the electromagnetic environment. However, several techniques have been developed to reduce this noise and improve the coherence times of superconducting qubits. One such technique is the use of pulse control methods, where tailored microwave pulses are applied to qubits to cancel out unwanted interactions and mitigate noise. These pulses are designed to correct errors caused by environmental disturbances and to preserve the coherence of the qubit during computations or measurements. In addition, superconducting qubits can be coupled to resonators, which act as filters to block specific noise frequencies while allowing the desired signal to pass through. This approach helps to isolate the qubit from environmental noise, improving its performance and stability in practical sensing applications.

7.3.7 Advances in Materials and Fabrication Techniques

Another important strategy for overcoming decoherence and noise is the development of new materials and fabrication techniques. The quality of the materials used in quantum devices plays a crucial role in their performance. For example, defects in the material can act as sources of noise, leading to decoherence and loss of quantum information.

Researchers are focusing on developing new materials with fewer defects and better stability for use in quantum sensors and computers. One such material is diamond, particularly the nitrogen-vacancy (NV) center, which has emerged as a promising candidate for quantum sensing. NV centers are stable at room temperature and exhibit long coherence times, making them ideal for applications in magnetometry, thermometry, and other types of quantum sensing. The development of high-quality diamond materials with minimal defects has significantly improved the performance of NV-based quantum sensors, demonstrating that advancements in material science are essential for overcoming the challenges of decoherence and noise.

7.4 QUANTUM SENSOR CALIBRATION TECHNIQUES

Quantum sensors have shown immense potential in providing unprecedented precision in measurements, revolutionizing fields such as metrology, navigation, imaging, and fundamental physics. Unlike classical sensors, which rely on classical properties of materials, quantum sensors exploit quantum properties such as superposition, entanglement, and interference to achieve higher sensitivity and accuracy. However, to ensure that these quantum sensors can provide reliable and reproducible results, calibration is essential. Calibration is the process of adjusting a sensor's measurements to match a known standard, ensuring that its output corresponds to the true value of the physical quantity being measured.

Quantum sensor calibration is a critical aspect of ensuring the accuracy and reliability of quantum measurements. Given the unique characteristics of quantum systems, calibration techniques for quantum sensors must account for the inherent sensitivity of quantum states to environmental factors and noise. Techniques such as self-calibration, quantum state tomography, and quantum interferometry are valuable tools for calibrating quantum sensors, each offering distinct advantages and challenges. Noise reduction and error correction are also essential for improving the accuracy of quantum sensor calibration. As quantum technology continues to advance, new calibration methods will be developed, enabling quantum sensors to reach their full potential in applications ranging from fundamental physics to medical diagnostics and environmental monitoring.

7.4.1 Challenges in Quantum Sensor Calibration

The calibration of quantum sensors is inherently more complex than that of classical sensors due to the fragile nature of quantum states. Quantum sensors often rely on quantum coherence, which is easily disrupted by environmental factors such as noise, temperature fluctuations, and electromagnetic interference. Furthermore, quantum systems are often highly sensitive to the initialization and measurement protocols, and small errors can lead to large discrepancies in the results. As a result, developing effective calibration techniques for quantum sensors is a nontrivial task. Additionally, many quantum sensors operate in regimes where traditional calibration methods are not directly applicable. For instance, quantum interferometers, which rely on the superposition of quantum states, cannot be easily calibrated using conventional reference standards. This requires the development of new calibration methodologies that can account for the unique properties of quantum systems.

7.4.2 Absolute Calibration vs. Relative Calibration

Quantum sensor calibration techniques can generally be categorized into two types: absolute calibration and relative calibration. Absolute calibration involves comparing the quantum sensor's measurements to a well-established standard or reference. This type of calibration is necessary when the goal is to determine the true value of a physical quantity, such as the magnetic field strength or temperature, using a quantum sensor. Absolute calibration is often challenging because it requires a known reference, which may not always be available in the context of quantum measurements. In contrast, relative calibration involves comparing the measurements of the quantum sensor to those of another sensor or system under controlled conditions. This approach is often more practical when an absolute standard is not accessible or when the goal is to characterize the sensor's performance over time. For example, a quantum sensor might be calibrated relative to a classical reference sensor that is known to be accurate or stable. This method is commonly used when calibrating sensors like atomic clocks or quantum gravimeters, where precise reference points are often difficult to obtain.

7.4.3 Self-Calibration Methods

Self-calibration is an innovative approach to quantum sensor calibration that leverages the sensor's own internal properties or known system behavior. In this method, the quantum sensor can be designed in such a way that it can automatically calibrate itself during operation, often without the need for an external reference. This approach is particularly useful in applications where external calibration standards may not be readily available or where continuous calibration is required. A common example of self-calibration is in quantum sensing devices that use entanglement or interference to measure physical quantities. These devices can be designed to measure the expected interference patterns or entanglement correlations under specific conditions, allowing the sensor to adjust its internal parameters to correct any deviations from the expected outcomes. This self-calibration capability can help maintain the accuracy and precision of quantum sensors in long-term or real-time applications, such as quantum-based magnetometers or accelerometers.

7.4.4 Quantum State Tomography in Calibration

Quantum state tomography is a technique used to characterize and calibrate quantum sensors by fully reconstructing the quantum state of the system. It involves performing a series of measurements on the system in different bases and using this information to infer the state of the system. Quantum state tomography is particularly useful for sensors based on qubits, where the quantum state needs to be accurately known for proper calibration. The key advantage of quantum state tomography in calibration is its ability to provide a detailed and comprehensive view of the sensor's behavior. By measuring the sensor's state in various configurations and comparing the results to theoretical predictions, one can identify and correct calibration errors. Quantum state tomography is widely used in quantum computing and quantum communication systems, and its application in quantum sensor calibration is an emerging area of research. However, quantum state tomography is resource-intensive, as it typically requires many measurements and computational resources to reconstruct the quantum state. Nevertheless, this technique remains one of the most accurate methods for understanding the performance of quantum sensors and can be particularly useful for sensors that require high levels of precision.

7.4.5 Calibration Using Quantum Interferometry

Quantum interferometry is a technique that exploits the interference of quantum states to make precise measurements. Interferometric techniques are used in a wide range of quantum sensors, including those used for gravitational wave detection, magnetometry, and atomic clocks. The calibration of quantum interferometers requires careful control over the quantum states involved in the interference process, as well as accurate measurement of the interference patterns. One of the challenges in calibrating quantum interferometers is the sensitivity of the interference pattern to changes in the phase of the quantum states. Small changes in the environment, such as temperature or electromagnetic fluctuations, can significantly affect the interference pattern and, by extension, the sensor's output.

To calibrate a quantum interferometer, the system must be carefully aligned, and its sensitivity to external disturbances must be characterized. This can be done by applying known reference signals and adjusting the system parameters to ensure the output is accurate.

7.4.6 Noise Reduction and Error Correction in Calibration

Noise and errors are inherent in any quantum measurement process, and they can have a significant impact on the accuracy of quantum sensor calibration. To address this issue, several techniques for noise reduction and error correction have been developed. These techniques aim to minimize the impact of environmental disturbances and improve the reliability of quantum sensor measurements. One widely used method for noise reduction is to isolate the quantum system from external noise sources, such as electromagnetic fields, vibrations, and temperature fluctuations. This is often achieved by placing the sensor in a controlled environment, such as a vacuum chamber or a cryogenic setup, where the effects of external disturbances can be minimized. Another approach to improving calibration accuracy is to employ quantum error correction protocols. These protocols involve encoding the quantum information in such a way that errors caused by noise can be detected and corrected. In the context of quantum sensor calibration, error correction can help maintain the accuracy of the measurements even in the presence of noise, ensuring that the calibration process produces reliable results.

7.4.7 Future Directions and Challenges

As quantum sensors continue to evolve, new calibration techniques will be required to keep pace with advances in technology. One of the key challenges is the scaling of calibration methods for increasingly complex quantum systems. As the size and complexity of quantum sensors increase, calibration techniques must be able to handle higher-dimensional quantum states and more intricate system interactions. Another challenge is the development of universal calibration standards for quantum sensors. While classical sensors can often be calibrated using well-established reference standards, quantum sensors require new approaches due to their reliance on quantum properties. The establishment of global calibration standards for quantum devices will be a significant milestone in the field, ensuring that quantum sensors can be used reliably across different industries and applications.

7.5 ENHANCING QUANTUM SENSOR PERFORMANCE WITH ENTANGLEMENT

Quantum sensors are devices that exploit the principles of quantum mechanics to measure physical quantities with precision that far exceeds the capabilities of classical sensors. These sensors harness phenomena such as superposition, interference, and entanglement to achieve remarkable accuracy in measuring parameters like magnetic fields, temperature, acceleration, and time. While quantum sensors have already demonstrated impressive sensitivity, there is a continuous effort to enhance their performance further, particularly in the face of challenges such as noise and decoherence. One of the most promising avenues for improving quantum sensor performance is through the use of quantum entanglement. Entanglement, one of the most fascinating and mysterious phenomena in quantum mechanics, occurs when two or more particles become linked in such a way that the state of one particle is directly related to the state of another, regardless of the distance between them.

This unique property has profound implications for enhancing the sensitivity and accuracy of quantum sensors. Enhancing quantum sensor performance with entanglement has opened up exciting possibilities for achieving ultra-precise measurements in a variety of fields. By exploiting the correlations between entangled particles, quantum sensors can surpass classical limits, offering superior sensitivity and noise resilience. Applications such as gravitational wave detection, quantum magnetometry, accelerometry, and atomic clocks have all benefited from the use of entanglement, making them indispensable tools in science and technology. As research in quantum entanglement and quantum sensor technology continues to evolve, the future promises even greater advancements in the precision and versatility of quantum sensors, with potential impacts ranging from fundamental physics to medical diagnostics and beyond.

7.5.1 Quantum Entanglement in Quantum Sensors

In quantum sensors, the core principle of operation revolves around measuring the effects of physical quantities on the quantum states of the system. Entanglement offers a significant advantage in this process by enabling the creation of correlated quantum states that are more sensitive to changes in the measured quantity than classical systems. When particles such as atoms, photons, or electrons are entangled, their combined quantum state exhibits correlations that allow for more precise measurements of external factors like magnetic fields, gravitational forces, or temperature. Entanglement can be used in various ways to improve sensor performance. For example, entangled states can be employed to reduce measurement uncertainty beyond the limits imposed by classical systems. This improvement in sensitivity is crucial for applications such as gravitational wave detection, where the measurement precision required is extremely high. Entangled states are particularly useful in quantum interferometry, a technique that relies on the interference of quantum states to measure small phase shifts caused by the physical quantity being sensed.

7.5.2 Quantum Metrology and the Role of Entanglement

Quantum metrology is a field that focuses on the use of quantum systems to measure physical quantities with unprecedented accuracy. The sensitivity of a quantum sensor is often limited by the quantum uncertainty principle, which places fundamental bounds on the precision with which certain pairs of physical properties (such as position and momentum, or time and energy) can be simultaneously known. This limit, known as the standard quantum limit (SQL), represents the best possible precision achievable with classical sensors. However, when entanglement is introduced into the system, it allows quantum sensors to surpass the standard quantum limit and achieve greater sensitivity.

This improvement is due to the enhanced correlations between particles in an entangled state, which enable the sensor to respond more acutely to small changes in the measured quantity. By utilizing entangled states, quantum sensors can achieve what is known as the Heisenberg limit, a much stricter bound on measurement uncertainty that provides superior performance compared to classical methods. Entanglement also plays a crucial role in enhancing quantum metrology techniques such as quantum interferometry, which is widely used in gravitational wave detectors like LIGO. In these systems, entangled photons are used to measure extremely small phase shifts that correspond to the presence of gravitational waves. The entanglement of photons allows for the detection of these tiny shifts with much higher precision than would be possible using classical light sources.

7.5.3 Entanglement and Noise Reduction in Quantum Sensors

One of the primary challenges faced by quantum sensors is the presence of noise and decoherence. These factors can disrupt the delicate quantum states that the sensors rely on, causing a loss of precision in measurements. Environmental factors such as temperature fluctuations, electromagnetic interference, and vibrations can induce errors in the measurement process, limiting the overall performance of the sensor. Entanglement can help mitigate these effects by enabling quantum error correction techniques and enhancing the robustness of the sensor to noise. Specifically, entangled quantum systems can exhibit a property known as "quantum discord," which allows for better resilience against certain types of noise. Quantum discord is a measure of the total amount of quantum correlations present in a system, and when used strategically, it can enhance the noise tolerance of quantum sensors. For instance, in quantum magnetometers, entangled particles can be used to improve the signal-to-noise ratio, allowing for more accurate magnetic field measurements in the presence of noise. By leveraging entanglement, quantum sensors can maintain high sensitivity even when external disturbances are present. This ability to reduce the impact of noise is particularly valuable in applications where the measured signals are faint and prone to interference, such as in deep-space measurements or in detecting weak magnetic fields in biological systems.

7.5.4 Applications of Entanglement in Quantum Sensing

The application of entanglement to quantum sensing has led to significant advancements in a variety of fields, particularly those that require extremely precise measurements. Some notable applications include quantum-enhanced gravitational wave detection, quantum magnetometry, quantum accelerometry, and atomic clocks.

- **Gravitational Wave Detection:** Gravitational waves are ripples in spacetime caused by massive cosmic events such as the collision of black holes. Detecting these waves requires highly sensitive instruments capable of measuring incredibly small distortions in spacetime. Quantum entanglement has been used in the LIGO (Laser Interferometer Gravitational-Wave Observatory) experiment to enhance the sensitivity of interferometers, allowing for the detection of gravitational waves from distant cosmic events. Entangled photons are employed to reduce the noise and improve the precision of the interferometric measurements, pushing the limits of what is possible in gravitational wave detection.
- **Quantum Magnetometry:** Quantum sensors based on entanglement are also used in magnetometry to measure weak magnetic fields with high precision. These sensors exploit the quantum properties of particles such as electrons or atoms, utilizing entanglement to improve sensitivity beyond classical limits. Quantum magnetometers are employed in a range of applications, from detecting magnetic fields in the human brain for medical purposes to studying the Earth's magnetic field for geophysical research.
- **Quantum Accelerometry:** Quantum accelerometers, which are used to measure acceleration and gravitational forces, can also benefit from entanglement. By entangling particles within an accelerometer, the device can achieve better precision in measuring small changes in acceleration, which is crucial for applications in inertial navigation and geophysics. Entangled quantum states allow these sensors to surpass the limitations imposed by classical accelerometers, making them more useful in environments such as space exploration and underground mapping.
- **Atomic Clocks:** Atomic clocks are the most accurate timekeeping devices in existence, and their performance can be enhanced with quantum

entanglement. By entangling atoms, atomic clocks can achieve higher precision in measuring time intervals, which is essential for applications such as GPS systems, telecommunications, and fundamental physics experiments.

7.5.5 Challenges and Future Directions

While the use of entanglement in quantum sensing has shown significant promise, there are still challenges to be addressed. One of the key obstacles is the issue of decoherence, which causes entangled states to lose their quantum coherence over time. This phenomenon can limit the effectiveness of entangled quantum sensors, especially in real-world environments where noise and temperature fluctuations are prevalent. Researchers are exploring various techniques, such as error correction codes and advanced isolation methods, to mitigate the effects of decoherence. Another challenge lies in the scalability of entangled quantum systems. Creating and maintaining entangled states with large numbers of particles is a difficult task, and scaling up these systems for practical use in large-scale sensing applications remains an ongoing area of research.

7.6 QUANTUM-ENHANCED SENSING BEYOND THE STANDARD QUANTUM LIMIT

Quantum sensing is a rapidly developing field that exploits the principles of quantum mechanics to achieve unprecedented sensitivity and precision in measuring physical quantities. Unlike classical sensors, which are subject to inherent limitations imposed by classical physics, quantum sensors harness phenomena such as superposition, interference, and entanglement to push the boundaries of measurement capabilities. A key challenge in the development of quantum sensors has been the issue of sensitivity limits, which, for many sensors, are constrained by the so-called Standard Quantum Limit (SQL). The Standard Quantum Limit arises from the quantum nature of the measurement process itself, where there is a trade-off between precision and the noise inherent in the system. This limitation can be understood in terms of the Heisenberg uncertainty principle, which states that there are fundamental bounds on how precisely certain pairs of properties (such as position and momentum, or time and energy) can be simultaneously measured. While the SQL sets a practical limit for many traditional quantum sensors, recent advances in quantum technology have paved the way for quantum-enhanced sensing techniques that allow measurements beyond this limit. These advancements have opened new possibilities for precision measurement in a wide array of applications. Quantum-enhanced sensing beyond the Standard Quantum Limit represents a significant leap forward in our ability to measure physical quantities with unprecedented precision. By exploiting advanced quantum properties like entanglement and squeezed states, these sensors allow for measurements that surpass the classical limitations imposed by the Standard Quantum Limit. The potential applications of quantum-enhanced sensing span a wide range of fields, from fundamental physics to practical technologies. While challenges remain, the continued development of quantum-enhanced sensors will undoubtedly lead to breakthroughs in science and technology, enabling us to measure the world around us with greater accuracy than ever before.

7.6.1 The Standard Quantum Limit and Its Implications

To understand how quantum-enhanced sensing works, it is important to first recognize the significance of the Standard Quantum Limit. In quantum mechanics, the uncertainty principle dictates that measurements of certain quantities are subject to a fundamental trade-off. For example, in interferometric measurements, where a quantum sensor measures small changes in phase or displacement, the SQL represents the minimum uncertainty in the measurement. This uncertainty is inherent in the quantum state of the system, meaning that the precision with which a quantity can be measured is limited by the properties of the quantum state itself. The SQL is often described as the "classical limit" for quantum sensors, as it represents the best possible performance achievable with typical quantum states, such as coherent states or squeezed states. While these sensors can be highly sensitive, they are still bound by the SQL, which can limit their utility in applications that require extremely precise measurements. The SQL provides a baseline for the performance of quantum sensors, but it is not the ultimate limit. The real challenge in quantum sensing is how to surpass this limit and achieve higher precision.

7.6.2 Quantum-enhanced Sensing Beyond the Standard Quantum Limit

Quantum-enhanced sensing techniques aim to overcome the constraints of the Standard Quantum Limit by exploiting advanced quantum properties. One of the most effective methods for surpassing the SQL is the use of squeezed states of light or matter. Squeezed states are special quantum states that reduce the uncertainty in one property (such as phase or momentum) at the expense of increasing uncertainty in the conjugate property. By carefully manipulating squeezed states, it is possible to achieve a reduction in measurement noise, which translates into greater sensitivity beyond the SQL. Another approach to overcoming the SQL involves the use of entangled quantum states. Entanglement, a phenomenon where the states of two or more particles become correlated in such a way that the state of one particle depends on the state of another, has been shown to provide a significant advantage in quantum sensing. When used in interferometry, entangled states can enhance the sensitivity of measurements by allowing the system to detect smaller changes in the measured quantity. This process, referred to as quantum metrology, enables measurements to exceed the SQL, often reaching what is known as the Heisenberg limit. Entanglement in quantum sensing has proven to be particularly powerful in applications where extreme precision is required, such as gravitational wave detection, atomic clocks, and magnetometry. In these cases, the use of entangled photons or atoms allows for the detection of tiny variations in physical quantities with a level of accuracy that would be impossible using classical techniques. The ability to surpass the SQL is crucial for the advancement of technologies that rely on precision measurement.

7.6.3 Applications of Quantum-enhanced Sensing Beyond the SQL

Quantum-enhanced sensors that operate beyond the Standard Quantum Limit have a wide range of potential applications in science and technology. These applications span multiple fields, from fundamental physics to practical technologies that can improve daily life.

- **Gravitational Wave Detection:** The detection of gravitational waves, ripples in spacetime caused by massive cosmic events, requires the ability to measure extremely small distortions in spacetime. Interferometric gravitational wave detectors like LIGO rely on quantum-enhanced sensors to detect these minute changes in length. By using squeezed states of light in the interferometer arms, researchers have been able to surpass the SQL and improve the sensitivity of these detectors, enabling the detection of gravitational waves from distant astrophysical events.
- **Quantum Magnetometry:** Quantum magnetometers are devices used to measure magnetic fields with high precision. These sensors take advantage of the quantum properties of atoms or photons, using entanglement or squeezed states to achieve greater sensitivity than classical magnetometers. Quantum-enhanced magnetometers are used in a wide variety of applications, including medical imaging (e.g., magnetoencephalography), geophysical exploration, and materials science.
- **Atomic Clocks:** Atomic clocks are the most accurate timekeeping devices known, and quantum-enhanced atomic clocks are even more precise than their classical counterparts. By employing squeezed states of light, researchers have been able to reduce the uncertainty in time measurements, pushing the performance of atomic clocks beyond the SQL. These ultra-precise clocks have applications in GPS systems, telecommunications, and fundamental tests of physics.

- **Quantum Interferometry:** Quantum interferometers are used to measure small changes in phase or position with incredible sensitivity. By using entangled or squeezed quantum states in the interferometric setup, these devices can surpass the Standard Quantum Limit, allowing for the detection of minute phase shifts that would be undetectable using classical methods. Quantum interferometry has applications in precision measurements of displacement, gravitational fields, and even fundamental constants of nature.
- **Quantum Imaging:** Quantum-enhanced imaging techniques utilize quantum states of light, such as squeezed light, to improve the resolution and sensitivity of imaging systems. These advancements enable the detection of faint signals and small features that would otherwise be obscured by noise. Quantum imaging techniques have applications in areas such as biological imaging, semiconductor inspection, and astronomical observations.

7.6.4 Challenges and Limitations of Quantum-enhanced Sensing

While quantum-enhanced sensing techniques show great promise in overcoming the Standard Quantum Limit, there are still several challenges and limitations that need to be addressed. One of the primary challenges is the issue of decoherence, which refers to the loss of quantum coherence due to interactions with the environment. Decoherence can cause quantum states to collapse into classical states, effectively destroying the advantages offered by quantum mechanics. Researchers are working on methods to reduce decoherence, such as using error correction techniques and isolating quantum sensors from environmental noise.

Another challenge is the scalability of quantum-enhanced sensing technologies. Many of the techniques used to achieve enhanced precision, such as squeezed states and entanglement, require complex setups and careful control of quantum systems. Scaling these technologies for use in real-world applications requires significant advancements in engineering and materials science. Furthermore, the long-term stability and reliability of quantum-enhanced sensors must be ensured for them to be practical in everyday use.

7.6.5 Future Directions in Quantum-enhanced Sensing

The future of quantum-enhanced sensing is incredibly promising, with the potential to revolutionize a wide range of industries and scientific fields. Continued advancements in quantum technologies, such as the development of more robust and scalable quantum sensors, will enable even greater precision in measurements. Researchers are exploring new quantum states and techniques, such as topological states of matter and quantum feedback, to further push the boundaries of quantum-enhanced sensing. The integration of quantum sensors with classical technologies is also an exciting avenue for future development. For example, hybrid systems that combine quantum-enhanced sensors with classical sensor networks could lead to new applications in fields such as autonomous navigation, environmental monitoring, and healthcare.

CHAPTER 8

Challenges and Limitations of Quantum Sensors

8.1 DECOHERENCE AND ENVIRONMENTAL INTERFERENCE

In the realm of quantum mechanics, the study of decoherence and environmental interference is central to understanding the practical challenges of working with quantum systems. Quantum systems, by their nature, exhibit superposition and entanglement, allowing them to exist in multiple states simultaneously and display behaviors that defy classical expectations. These unique properties, however, are highly sensitive to external factors. When a quantum system interacts with its environment, it can lose its quantum coherence, a process known as decoherence. This loss of coherence is one of the major obstacles in the development of practical quantum technologies, such as quantum computing, quantum sensors, and quantum communication systems. Decoherence and environmental interference present significant obstacles in the development of practical quantum technologies. As quantum systems interact with their environment, they lose their quantum coherence, causing them to behave more classically and limiting their performance. Understanding the mechanisms behind decoherence and finding ways to mitigate its effects are crucial for the advancement of quantum technologies. Techniques such as quantum error correction, dynamical decoupling, and the use of decoherence-free subspaces are essential for reducing the impact of environmental interference on quantum systems. By continuing to improve our understanding and control of decoherence, the potential for quantum technologies in fields such as quantum computing, quantum sensing, and quantum communication can be fully realized.

8.1.1 Understanding Decoherence

Decoherence refers to the process by which a quantum system loses its ability to exhibit quantum behaviors due to interactions with its environment. In an ideal isolated quantum system, particles can exist in a superposition of multiple states, leading to interference effects and the ability to process information in parallel. However, when such a system interacts with its environment, the system's quantum state becomes entangled with the states of the surrounding environment. This entanglement effectively "measures" the quantum system, causing the superposition to collapse into a definite state. As a result, the system appears to behave more classically, and the quantum coherence (the ability to maintain superposition) is lost. The process of decoherence can be understood as a transition from a pure quantum state to a mixed state. In a pure state, the system has well-defined probabilities for each outcome, and interference effects can occur. In a mixed state, these interference effects disappear because the system is no longer in a well-defined quantum state, but rather in a probabilistic mixture of states. Decoherence does not necessarily destroy the information in the system, but it destroys the quantum correlations and the delicate interference patterns that are central to quantum mechanical behavior.

8.1.2 Environmental Interference: The Role of the Environment

Environmental interference plays a crucial role in the process of decoherence. The environment, in this context, refers to all external factors—such as thermal fluctuations, electromagnetic fields, or even cosmic radiation—that a quantum system interacts with. These factors cause the system to become entangled with the environment, leading to decoherence. The environment essentially acts as a "measuring device" that monitors the quantum system, forcing it to choose a definite state. The strength and nature of the environmental interactions determine how quickly decoherence occurs. For example, at higher temperatures, the interactions between the quantum system and its environment become more frequent, leading to faster decoherence. In contrast, a cold, isolated system may experience less interference, allowing the quantum state to persist for longer periods before decohering. In addition to thermal effects, electromagnetic noise, such as fluctuations in electric and magnetic fields, can also contribute to decoherence. For instance, a quantum computer's qubits are highly susceptible to such environmental noise, which can cause errors in the computation. The challenge lies in mitigating these environmental factors to preserve the coherence of quantum states for longer periods, which is essential for the functioning of quantum devices.

8.1.3 Types of Decoherence

There are several different types of decoherence that can occur depending on the nature of the interaction between the quantum system and its environment. One of the most common types is phase decoherence, which occurs when the relative phases between components of a quantum superposition are disrupted. This type of decoherence is particularly important in quantum computing and quantum sensors, where maintaining the phase coherence of quantum states is crucial for the operation of quantum algorithms or measurements. Another form of decoherence is dephasing, which occurs when the quantum system's energy levels become coupled to the environment, causing random fluctuations in the system's phase. Dephasing can result from interactions with the electromagnetic field, for example, and can significantly degrade the performance of quantum systems, particularly those used in quantum information processing. In addition to phase decoherence and dephasing, bit-flip and phase-flip decoherence also occur, especially in quantum computing contexts. A bit-flip happens when the quantum state of a system flips between two computational states (e.g., from 0 to 1), while a phase-flip alters the phase of the quantum state without changing its computational basis. These types of decoherence disrupt the functioning of quantum algorithms, causing errors and loss of information.

8.1.4 Factors Contributing to Decoherence

The rate and nature of decoherence are influenced by several factors. The strength of the coupling between the quantum system and its environment is one of the most significant factors. Stronger coupling leads to faster decoherence, as the quantum system becomes increasingly entangled with the environment. Conversely, weaker coupling allows the system to retain its coherence for longer periods, which is advantageous for quantum technologies. Another key factor is the temperature of the environment. Higher temperatures increase the level of thermal noise, which accelerates decoherence. At low temperatures, quantum systems can be better isolated from the environment, reducing the risk of decoherence and enabling quantum states to be maintained for longer periods. For instance, many experiments involving quantum computing use cryogenic temperatures to reduce thermal noise and slow down decoherence. The physical environment of the quantum system also plays a crucial role. Electromagnetic fields, vibrations, and even cosmic radiation can interact with a quantum system, leading to decoherence. Shielding quantum systems from these external influences is a significant challenge in building practical quantum technologies. For example, quantum computers often require sophisticated error correction codes to mitigate the effects of decoherence caused by environmental interference.

8.1.5 Mitigating Decoherence and Environmental Interference

To fully exploit the potential of quantum systems in practical applications, it is necessary to find ways to mitigate the effects of decoherence and environmental interference. Several strategies are currently being developed to address these challenges. One approach is the use of quantum error correction techniques, which aim to detect and correct errors caused by decoherence before they affect the computation or measurement. Quantum error correction relies on encoding information in a way that makes it resilient to noise and decoherence. While error correction can help reduce the impact of environmental interference, it also requires additional resources, such as extra qubits or computational power, which can be a limitation in certain applications. Another strategy is the development of decoherence-free subspaces. In this approach, quantum systems are engineered to interact in such a way that certain degrees of freedom are unaffected by the environment. This technique can help protect quantum systems from certain types of decoherence, enabling them to maintain coherence for longer periods.

8.1.6 Quantum Control and Dynamical Decoupling

In addition to error correction and decoherence-free subspaces, quantum control techniques such as dynamical decoupling are being explored to combat decoherence. Dynamical decoupling involves applying a sequence of carefully timed operations to the quantum system in order to counteract the effects of environmental noise. By "resetting" the system at regular intervals, it is possible to prevent the buildup of decoherence, allowing quantum coherence to persist for longer periods. These techniques have shown promise in enhancing the stability and reliability of quantum sensors and quantum computers.

8.1.7 The Future of Decoherence Mitigation

As the field of quantum technologies continues to advance, new methods for mitigating decoherence and environmental interference are likely to emerge. Researchers are exploring the use of exotic materials and quantum error correction codes that can further reduce the impact of noise on quantum systems. Additionally, advances in quantum control and the development of novel quantum states may provide new ways to maintain coherence in challenging environments, such as in the presence of strong electromagnetic fields or at elevated temperatures. The development of practical quantum systems that are less susceptible to decoherence is crucial for the realization of robust quantum computers, quantum sensors, and other quantum technologies. As our understanding of decoherence and environmental interference deepens, it will be possible to develop more effective strategies to combat these challenges and unlock the full potential of quantum mechanics.

8.2 TECHNOLOGICAL CHALLENGES IN SCALING QUANTUM SENSORS

Quantum sensors represent a leap forward in precision measurement, offering unparalleled sensitivity and accuracy in a variety of fields, from fundamental physics to medical diagnostics. By exploiting quantum properties such as superposition and entanglement, these sensors are capable of detecting physical quantities with a precision that far exceeds that of classical sensors. However, scaling quantum sensors for practical, widespread use presents significant technological challenges. These challenges range from the fundamental limitations imposed by quantum mechanics to the practical difficulties in creating systems that are robust, cost-effective, and scalable. As researchers continue to push the boundaries of quantum sensing, understanding and addressing these technological hurdles is crucial for realizing the potential of quantum sensors in real-world applications. The scaling of quantum sensors presents a range of technological challenges that must be addressed to unlock their full potential. These challenges include maintaining quantum coherence over long periods, overcoming material limitations, scaling quantum systems, and dealing with cryogenic requirements and environmental interference. As quantum sensing technologies continue to evolve, innovations in materials science, quantum control techniques, and cryogenic engineering will be key to addressing these challenges. By overcoming these hurdles, quantum sensors have the potential to revolutionize fields such as navigation, medical diagnostics, and fundamental physics, bringing their extraordinary sensitivity and precision to practical, real-world applications.

8.2.1 Quantum Coherence and Decoherence

One of the primary technological challenges in scaling quantum sensors is maintaining quantum coherence over long periods. Quantum systems, such as qubits or other quantum states used in sensing, rely on maintaining a delicate balance of superposition, where the system can exist in multiple states at once. This quantum coherence is essential for the precise measurement capabilities of quantum sensors. However, quantum coherence is highly fragile and can easily be disrupted by environmental factors, such as electromagnetic fields, temperature fluctuations, and even cosmic radiation. This disruption, known as decoherence, causes the quantum system to lose its ability to maintain a superposition state, thereby degrading the sensor's performance. For practical quantum sensors to be scaled and deployed in real-world environments, researchers must find ways to mitigate the effects of decoherence. This requires developing advanced isolation techniques to shield quantum systems from environmental interference, as well as creating error-correction protocols that can maintain the integrity of quantum states. While these solutions are promising, they add complexity to the design and operation of quantum sensors, limiting their scalability in some cases.

8.2.2 Material Limitations and Fabrication Challenges

Another significant hurdle in scaling quantum sensors is the limitation of available materials. Quantum sensors often rely on specific materials that exhibit quantum properties, such as superconducting materials, optically active defects, or trapped ions. These materials must be carefully engineered to exhibit the desired quantum behavior and be stable under the conditions required for sensing. However, scaling these materials to a level where they can be used in large, practical devices is no simple task. For example, certain quantum sensors, such as those based on nitrogen-vacancy (NV) centers in diamond, require high-quality diamond crystals with precisely positioned defects. Fabricating these materials with the necessary purity and precision is challenging, and it becomes even more difficult when attempting to scale up production for widespread use. Similarly, superconducting sensors such as SQUIDs (superconducting quantum interference devices) require extremely low temperatures to function effectively, which necessitates complex and costly cryogenic cooling systems. As quantum sensing technology progresses, there is a need for innovations in materials science and fabrication techniques that allow for the production of large-scale, high-performance quantum sensors. This could involve discovering new materials with better quantum properties or developing more efficient methods for creating and controlling quantum states in materials.

8.2.3 Scalability of Quantum Systems

A fundamental challenge in scaling quantum sensors lies in the inherent difficulty of scaling up quantum systems themselves. Quantum sensors often rely on small, controlled quantum systems, such as a few qubits or atoms, to perform precise measurements. However, as the size of the quantum system increases, the complexity of maintaining quantum coherence and controlling the system also increases exponentially. For example, in quantum computers, scaling up the number of qubits while maintaining their coherence is a well-known challenge. Similarly, in quantum sensors, scaling up the number of sensing elements can lead to increased noise and interference, reducing the overall precision and performance of the sensor. To overcome this, researchers are exploring novel techniques such as quantum error correction and the development of quantum systems that are less prone to environmental disturbances. Additionally, the integration of quantum sensors into practical devices requires that the sensors be miniaturized and packaged in a way that allows them to operate effectively in real-world settings. This miniaturization often involves complex engineering challenges, including the need for precise alignment, integration of cooling systems, and ensuring reliable performance over long periods of use.

8.2.4 Cryogenic Requirements and Temperature Sensitivity

Many quantum sensors, particularly those based on superconducting technologies or cold atom sensors, require extremely low temperatures to operate. This necessitates the use of cryogenic cooling systems, which are expensive, bulky, and energy-intensive. Cryogenic cooling is essential to reduce thermal noise and prevent decoherence in certain quantum systems, but it also limits the scalability and practicality of quantum sensors, particularly in mobile or field applications. The challenge of maintaining cryogenic temperatures over long periods is exacerbated by the need for complex infrastructure, such as cryostats or dilution refrigerators, which can be both costly and difficult to maintain. Additionally, the sensitivity of quantum sensors to temperature fluctuations means that even small changes in temperature can have a significant impact on sensor performance. As a result, scaling quantum sensors for use in practical settings will require innovations in cryogenic technology and more efficient cooling methods.

8.2.5 Noise and Environmental Interference

Quantum sensors are highly sensitive to external noise and environmental interference, which can significantly affect their performance. Sources of noise include electromagnetic interference, vibrations, acoustic noise, and even cosmic rays. Quantum systems, by their nature, are highly susceptible to such disturbances, which can cause decoherence, reduce precision, and introduce errors in measurements. Addressing environmental noise and interference is one of the most significant challenges in scaling quantum sensors.

For many quantum sensors, such as those based on nitrogen-vacancy centers in diamonds or optomechanical systems, shielding from external noise is essential. However, achieving effective shielding is difficult, particularly in complex, real-world environments where environmental factors can be highly variable. Additionally, noise can vary in frequency and intensity, making it challenging to develop shielding that works across a broad range of conditions. To mitigate these effects, researchers are developing techniques such as dynamical decoupling, where control pulses are applied to quantum systems to counteract the effects of noise. Other approaches involve using error-correction codes or filtering techniques to separate signal from noise. While these methods show promise, they can increase the complexity of the system and may require additional computational resources, which can be a barrier to scalability.

8.2.6 Cost and Practical Integration

While quantum sensors offer unparalleled precision, they often come with a high price tag due to the need for specialized materials, cooling systems, and advanced control mechanisms. The cost of manufacturing, maintaining, and operating quantum sensors is currently a significant barrier to their widespread adoption. In addition, many quantum sensors require highly specialized expertise to operate, which limits their accessibility to research institutions and specialized industries. To overcome this challenge, the development of cost-effective quantum sensors that can be produced on a larger scale is essential.

This involves creating new manufacturing techniques that reduce the complexity and cost of quantum sensor production, as well as developing more user-friendly systems that can be operated by non-experts. Advances in quantum technologies could lead to the mass production of quantum sensors, making them more affordable and accessible for a wide range of applications.

8.3 COST AND PRACTICALITY OF QUANTUM SENSING SYSTEMS

Quantum sensing systems promise to revolutionize various fields, including medical diagnostics, environmental monitoring, and fundamental physics research. Leveraging quantum phenomena such as superposition, entanglement, and coherence, these sensors offer extraordinary sensitivity and precision, often surpassing that of classical counterparts. However, while quantum sensors offer unprecedented capabilities, their cost and practicality remain significant barriers to their widespread adoption. The complexity of quantum technologies, along with the specialized materials and infrastructure required to support them, makes quantum sensors expensive and challenging to integrate into practical applications. Quantum sensors hold immense promise for a wide range of applications, from advancing our understanding of the universe to enabling breakthroughs in medical diagnostics and environmental monitoring. However, the cost and practicality of these systems remain significant challenges that must be addressed to make them viable for widespread use. As quantum sensing technologies continue to evolve, overcoming the barriers associated with high costs, complexity, and environmental requirements will be essential. Advances in materials science, manufacturing techniques, and operational efficiency will play a crucial role in reducing the cost of quantum sensors and making them more practical for everyday applications. By addressing these challenges, quantum sensors could become more affordable, accessible, and widely used, opening up new possibilities across numerous fields.

8.3.1 The High Costs of Quantum Sensing Components

The cost of quantum sensors primarily stems from the specialized components required to create and maintain quantum states. These components often include high-quality materials with specific quantum properties, such as superconducting materials for superconducting quantum interference devices (SQUIDs), nitrogen-vacancy (NV) centers in diamonds, and optically active materials for cold atom sensors. These materials must be carefully sourced, processed, and integrated into the sensor system, which increases production costs. In addition to the material costs, the fabrication process for quantum sensors is highly complex and requires advanced manufacturing techniques. For example, the precise positioning of defects in diamond to create NV centers or the use of ultra-high-vacuum environments for trapping and manipulating atoms adds significant expenses. Moreover, scaling these production processes to create larger or more efficient quantum sensors further drives up costs, as it requires advanced equipment and highly skilled labor.

8.3.2 Cryogenic Requirements and Their Impact on Cost

Many quantum sensing systems, particularly those based on superconducting materials or cold atom technologies, require extremely low temperatures to operate effectively. For example, SQUIDs, used for highly sensitive magnetic field detection, require cooling to temperatures close to absolute zero, necessitating the use of cryogenic refrigeration systems. These cooling systems are typically large, energy-intensive, and costly, adding another layer of expense to quantum sensor systems. In addition to the upfront cost of purchasing and maintaining cryogenic systems, the operational cost of keeping quantum sensors at these low temperatures can be substantial. The need for continuous refrigeration, along with the associated energy consumption and infrastructure maintenance, makes it impractical for many potential applications where portability and cost-effectiveness are crucial. As quantum sensing technologies continue to develop, reducing the reliance on cryogenic temperatures or finding more efficient ways to cool these systems will be essential for making quantum sensors more affordable and practical for real-world use.

8.3.3 Complexity and Expertise in Operation

Quantum sensing systems require a high level of technical expertise to operate. These systems are based on the manipulation and control of quantum states, which is inherently complex and requires precise calibration and monitoring. Additionally, quantum sensors are highly sensitive to environmental factors, such as temperature fluctuations, electromagnetic fields, and vibrations, which means they must be carefully shielded and maintained. This makes the operation of quantum sensors a specialized task, often requiring skilled scientists or engineers. The need for specialized knowledge to operate quantum sensors adds to their cost. Institutions or companies wishing to use quantum sensors may need to hire experts with extensive training in quantum mechanics, material science, and sensor technology. Furthermore, maintaining and troubleshooting these systems requires ongoing technical support, which further drives up operational costs. For quantum sensors to become widely accessible, the development of user-friendly systems that require less expertise to operate will be crucial. These systems must be designed in a way that allows for ease of integration into existing infrastructure and industries without compromising the sensor's performance.

8.3.4 Infrastructure and Environmental Control

In addition to the cost of the sensors themselves, the infrastructure required to support quantum sensors also contributes to their overall cost. Quantum sensors often need to be placed in carefully controlled environments to function properly. This can involve sophisticated vacuum chambers, magnetic shielding, and vibration isolation platforms to reduce the influence of environmental noise and maintain quantum coherence. For example, quantum sensors that rely on cold atom technologies, such as atom interferometers, require ultra-high-vacuum environments and magnetic field shielding to prevent decoherence. These requirements mean that quantum sensors cannot simply be placed in any environment; instead, they need specialized setups, which can be expensive to build and maintain.

The need for this infrastructure makes it difficult to scale quantum sensors for use in many real-world settings, particularly for applications requiring portability, such as field measurements or consumer-level devices. As the field of quantum sensing evolves, addressing these infrastructure needs will be essential. Developing portable, robust, and easy-to-assemble quantum sensors that do not require highly specialized environments will be key to reducing costs and making quantum sensing more practical for a wider range of applications.

8.3.5 Scaling Quantum Sensors for Mass Production

One of the significant challenges in reducing the cost of quantum sensors is scaling up production to meet demand. The current production of quantum sensors is often limited to small batches or highly specialized applications due to the complexity and cost of manufacturing. To bring down costs, it will be necessary to find ways to scale up the production process while maintaining the high precision and performance required for these sensors. Scaling production of quantum sensors involves addressing the challenges of mass-producing high-quality materials, integrating multiple quantum components into a single sensor, and ensuring consistency and reliability across large quantities of sensors. This requires innovations in manufacturing technologies, such as automated fabrication processes or new methods of synthesizing quantum materials, which can be produced at a lower cost without sacrificing performance.

Moreover, scaling quantum sensors also requires advancements in packaging and integration. As the size of quantum sensing systems decreases, their integration into existing devices and infrastructure will become more feasible. Packaging quantum sensors in a way that makes them easy to incorporate into a variety of applications—whether in medicine, industry, or research—will be a critical step in making quantum sensors more affordable and practical.

8.3.6 Reducing Operational Costs and Making Sensors Accessible

Aside from the initial high costs, the operational expenses associated with maintaining quantum sensors are also a significant barrier to widespread adoption. For example, the cryogenic cooling systems required for certain quantum sensors are energy-intensive and costly to operate continuously. Additionally, maintaining the environmental conditions necessary for optimal sensor performance can require significant infrastructure and ongoing technical support. Efforts to reduce operational costs must focus on improving the energy efficiency of quantum sensor systems, finding more cost-effective cooling methods, and minimizing the need for complex infrastructure. For instance, researchers are exploring the use of room-temperature quantum sensors or sensors that operate at higher temperatures, reducing the need for expensive and energy-hungry cryogenic cooling. These advancements would greatly reduce the operational costs associated with quantum sensors, making them more accessible to a wider range of industries and applications.

8.4 NOISE MANAGEMENT AND ERROR CORRECTION IN QUANTUM SENSORS

Quantum sensors exploit the unique properties of quantum mechanics to achieve unparalleled sensitivity and precision in measurement. However, the very characteristics that give quantum sensors their power, such as superposition and entanglement, also make them susceptible to noise and errors. Noise management and error correction are critical components in maximizing the performance of quantum sensors and ensuring their reliability in practical applications. The complexity of quantum systems introduces several challenges that can degrade the accuracy of measurements, and understanding how to control or mitigate these issues is vital for the successful deployment of quantum sensors in various fields. Noise management and error correction are essential components of quantum sensing systems, as they directly affect the accuracy and reliability of measurements. Quantum sensors, by their very nature, are susceptible to a range of noise sources, both classical and quantum, that can degrade their performance. However, through effective noise management techniques such as environmental shielding, robust system design, and operational control, the impact of noise can be minimized. In addition, quantum error correction plays a critical role in ensuring the stability of quantum sensors by protecting quantum information from errors caused by noise. Although quantum error correction presents significant challenges, ongoing advancements in error-correcting codes and machine learning techniques show great promise for improving the performance of quantum sensors. As quantum sensor technologies continue to evolve, the development of efficient noise management and error correction strategies will be crucial for making these sensors practical for widespread use across diverse fields. By overcoming the challenges associated with noise and errors, quantum sensors will be able to reach their full potential, providing unprecedented precision and sensitivity in a wide range of applications.

8.4.1 Sources of Noise in Quantum Sensors

Quantum sensors rely on the manipulation of quantum states, such as superposition and entanglement, to measure physical quantities. However, quantum systems are highly sensitive to external influences, leading to a variety of noise sources that can introduce errors in measurements. These noise sources can generally be categorized into two types: classical noise and quantum noise. Classical noise refers to the external disturbances in the environment, such as temperature fluctuations, vibrations, or electromagnetic interference, which can affect the system's performance. For instance, temperature fluctuations can induce changes in the properties of the materials used in quantum sensors, leading to drifts in measurements. Similarly, vibrations can perturb the positioning of atoms or ions used in sensing systems, causing fluctuations in the precision of measurements. Quantum noise, on the other hand, arises from the intrinsic uncertainties that govern quantum systems. This includes effects such as quantum fluctuations and decoherence. Quantum fluctuations refer to the random variations in the quantum state of a system, which can introduce noise into measurements. Decoherence is the process by which a quantum system loses its quantum coherence due to interaction with the environment, effectively causing the system to behave more classically. Decoherence and quantum fluctuations are particularly problematic in systems like quantum interferometers or atom interferometers, where precise phase measurements are required.

8.4.2 Impact of Noise on Quantum Sensors

Noise in quantum sensors directly impacts their sensitivity and accuracy. In many cases, noise can reduce the precision of measurements beyond the theoretical limits, rendering the sensor ineffective or unreliable for practical use. For example, in quantum gravimeters or magnetometers, noise can lead to inaccuracies in detecting weak gravitational fields or magnetic fields, limiting their utility in applications such as geophysical surveying or medical imaging. In the context of quantum interferometry, noise can cause phase shifts that distort the measurement signal, leading to errors in determining the phase difference between interfering quantum states. These errors can accumulate over time, degrading the sensor's overall performance. Similarly, in quantum thermometers or accelerometers, noise can introduce errors in temperature or acceleration measurements, reducing their precision and reliability. Additionally, the presence of noise can reduce the sensitivity of quantum sensors to weak signals. Quantum systems often work by amplifying small signals, and noise can overwhelm the system's ability to detect these signals. For example, in gravitational wave detection, the minute fluctuations in spacetime caused by distant astrophysical events require extremely sensitive sensors, and any noise could mask these signals, preventing accurate detection.

8.4.3 Strategies for Noise Management in Quantum Sensors

To address these challenges, effective noise management strategies are essential for improving the performance of quantum sensors. One of the key strategies is environmental shielding, which involves isolating quantum systems from external disturbances such as temperature fluctuations, electromagnetic fields, and vibrations. Shielding materials and techniques, such as superconducting magnetic shields or vibration isolation platforms, are commonly used in quantum sensing systems to reduce the impact of classical noise. Another critical strategy for noise management is the use of robust quantum systems that are less susceptible to external influences. For example, some quantum sensors utilize atomic or molecular systems that are inherently more stable and less sensitive to environmental perturbations. Using materials with low susceptibility to noise and designing systems that minimize interaction with the environment can enhance the resilience of quantum sensors. Additionally, some quantum sensors are designed to operate in isolated or controlled environments, such as ultra-high-vacuum chambers, to reduce the impact of external disturbances. The ability to operate in such controlled environments allows quantum sensors to achieve greater precision and accuracy by minimizing the noise that arises from interactions with the surrounding environment.

8.4.4 Error Correction Techniques in Quantum Sensors

In addition to noise management, error correction techniques are crucial for ensuring the accuracy and reliability of quantum sensors. Quantum error correction is a highly active area of research, as it aims to protect quantum information from the errors that arise due to noise and imperfections in quantum systems. Traditional error correction techniques, which work well for classical systems, do not apply directly to quantum systems due to the nature of quantum mechanics. Quantum error correction techniques exploit the properties of quantum entanglement and superposition to detect and correct errors in quantum states. One of the most well-known approaches is the use of quantum error-correcting codes, such as the Shor code or the surface code. These codes encode quantum information in a way that allows errors to be detected and corrected without measuring the quantum state directly, thus avoiding the collapse of the quantum wavefunction. For instance, in the Shor code, quantum information is encoded into multiple qubits, and redundancy is introduced to protect against errors. If an error occurs, it can be detected by comparing the states of different qubits and correcting the error through a process called "syndromic decoding." This technique ensures that even in the presence of noise, the quantum information remains intact, and the sensor continues to operate with high precision. Another error correction technique is known as decoherence-free subspaces, which are designed to protect quantum states from the effects of decoherence. By encoding quantum information into specific states that are immune to certain types of noise, these subspaces allow quantum sensors to maintain coherence over longer periods, thereby improving their stability and reducing the impact of environmental noise.

8.4.5 Improving Quantum Error Correction for Practical Use

While quantum error correction holds great promise, its practical implementation in quantum sensors faces several challenges. One of the main issues is the overhead associated with error correction. For example, quantum error-correcting codes often require additional qubits and computational resources, which can increase the complexity and cost of quantum sensors. Moreover, the decoding process can be time-consuming, leading to delays in obtaining measurement results. To address these challenges, researchers are developing more efficient error correction schemes that minimize the overhead and can be implemented in real-time. For example, techniques such as concatenated codes or fault-tolerant quantum computation are being explored to improve the efficiency of error correction in quantum sensors. Additionally, advances in hardware, such as the development of quantum processors with high-fidelity qubits, will reduce the error rates and the need for extensive error correction.

8.4.6 The Role of Machine Learning in Noise Management and Error Correction

An emerging approach to noise management and error correction in quantum sensors is the use of machine learning and artificial intelligence (AI) techniques. Machine learning algorithms can be trained to identify and mitigate noise in real-time by analyzing patterns in sensor data and predicting potential errors. These algorithms can be used to improve the sensitivity and precision of quantum sensors by detecting environmental disturbances and adjusting the sensor's parameters accordingly.

For example, machine learning algorithms can be used to optimize the calibration of quantum sensors by identifying patterns in measurement errors and automatically adjusting the sensor's settings to compensate for them. This approach can be particularly useful in environments with unpredictable or fluctuating noise, where traditional noise management strategies may be less effective. Furthermore, machine learning can be integrated with quantum error correction techniques to enhance their performance. By using AI to predict the likelihood of errors and provide feedback on error correction processes, machine learning can improve the efficiency and accuracy of quantum error correction in real-time, allowing quantum sensors to maintain their performance even in the presence of significant noise.

8.5 INTEGRATION OF QUANTUM SENSORS WITH CLASSICAL SYSTEMS

The integration of quantum sensors with classical systems represents a significant advancement in the field of precision measurement. Quantum sensors are renowned for their ability to exploit quantum mechanical phenomena, such as superposition, entanglement, and coherence, to achieve unprecedented sensitivity and accuracy in measuring physical quantities. However, their full potential can only be realized when they are effectively integrated with classical systems, which are widely used in many practical applications. The challenge lies in developing systems that bridge the gap between the unique capabilities of quantum sensors and the robust, scalable infrastructure of classical technologies.

This integration allows quantum sensors to complement and enhance existing classical systems, offering a broader range of applications and improved performance. The integration of quantum sensors with classical systems represents a promising avenue for enhancing the capabilities of current technologies. By combining the precision of quantum sensors with the reliability of classical systems, it is possible to achieve unprecedented sensitivity and accuracy in measurement, opening up new possibilities for a wide range of applications. As quantum sensing technology continues to evolve, the integration of quantum and classical systems will play a key role in shaping the future of sensing and measurement.

8.5.1 The Need for Integration

Quantum sensors, such as those based on atomic clocks, magnetometers, or quantum interferometers, have demonstrated remarkable performance in laboratory settings. However, these sensors often face limitations when it comes to real-world applications. One key challenge is the need for classical systems to provide the necessary infrastructure for quantum sensors to function in practical environments.

Classical systems are designed to handle large-scale operations, providing data processing, communication, and user interfaces, which are essential for quantum sensors to be deployed effectively in real-world scenarios. Another reason for integrating quantum sensors with classical systems is that many practical systems, including navigation, communication, and monitoring devices, are based on classical technologies. For example, GPS systems, which rely on classical satellite-based technology, are widely used for navigation. However, GPS signals can be blocked or degraded in certain environments, such as in deep urban canyons or underground. Quantum sensors, such as quantum gyroscopes, offer a potential solution by providing highly accurate inertial measurements without the need for GPS signals. Integrating these quantum sensors with classical systems can lead to more resilient and versatile navigation solutions.

8.5.2 Challenges in Integration

While the integration of quantum sensors with classical systems offers significant benefits, it also presents a number of challenges. One of the primary challenges is the inherent differences between quantum and classical systems. Quantum systems are highly sensitive to environmental disturbances, such as temperature fluctuations, vibrations, and electromagnetic interference. In contrast, classical systems are generally more stable and less sensitive to such disturbances. As a result, integrating quantum sensors with classical systems requires overcoming the challenges of noise, decoherence, and system instability. Another challenge is the complexity of controlling and managing quantum sensors. Quantum systems are governed by the principles of quantum mechanics, which are fundamentally different from the classical physics that governs most of the technologies used in everyday life. Quantum sensors require precise control over the quantum states of particles, which is a complex and delicate task. Classical systems, on the other hand, are generally easier to manage and control. Developing interfaces that allow classical systems to communicate with and control quantum sensors requires sophisticated technology and expertise. Additionally, quantum sensors often require highly specialized hardware and software, which can be difficult to integrate into existing classical systems. Classical systems are typically designed to work with conventional sensors that operate based on classical physics principles. Introducing quantum sensors into these systems may require significant modifications to hardware, software, and communication protocols.

8.5.3 Strategies for Successful Integration

Despite the challenges, several strategies can facilitate the successful integration of quantum sensors with classical systems. One of the most important strategies is the development of hybrid systems that combine the strengths of both quantum and classical technologies. Hybrid systems leverage the precision and sensitivity of quantum sensors while utilizing the robustness and scalability of classical systems. For example, quantum sensors can be used to provide highly accurate measurements, which can then be processed and interpreted by classical systems to deliver practical results. In this way, quantum sensors enhance the performance of classical systems without requiring a complete overhaul of existing infrastructure. In some cases, quantum sensors can be used in conjunction with classical sensors to create more comprehensive sensing systems. For instance, quantum sensors can complement classical sensors by providing high-precision measurements in environments where classical sensors may struggle. In a magnetic resonance imaging (MRI) system, for example, classical sensors can be used to detect general signals, while quantum sensors, such as nitrogen-vacancy (NV) centers in diamond, can provide more precise measurements of magnetic fields. Integrating these sensors allows for a more complete and accurate understanding of the system being measured. Another important strategy is the development of control systems that allow classical systems to manage and operate quantum sensors.

Quantum control is a complex field, and researchers are developing specialized algorithms and hardware to enable classical systems to interact with and control quantum sensors. These control systems can be used to monitor and correct errors in quantum sensors, ensuring that they remain stable and accurate in real-world environments. Furthermore, classical systems can be used to process the data obtained from quantum sensors, providing the necessary tools for data analysis, visualization, and decision-making.

8.5.4 Applications of Quantum Sensors in Classical Systems

The integration of quantum sensors with classical systems opens up a wide range of applications across various fields. One of the most promising areas is in navigation and positioning systems. Quantum gyroscopes, which are highly sensitive to rotation, can be integrated with classical accelerometers and magnetometers to provide accurate inertial navigation in environments where GPS signals are unavailable or unreliable. This combination of quantum and classical sensors is already being explored for use in autonomous vehicles, underwater navigation, and space exploration. Another area where quantum sensors can enhance classical systems is in the field of environmental monitoring.

Quantum sensors, such as those based on atomic interferometry or NV centers, can be used to detect minute changes in magnetic fields, temperature, and gravitational fields. By integrating these sensors with classical monitoring systems, it is possible to create highly sensitive and precise environmental sensors that can be used for a wide range of applications, including geophysical surveys, earthquake monitoring, and climate studies. In the field of healthcare, quantum sensors can be integrated with classical medical technologies, such as MRI or ultrasound, to provide more detailed and accurate diagnostic information. For example, NV centers in diamond can be used to measure the magnetic fields generated by biological tissues, providing insights into cellular processes or detecting early signs of disease. By combining these quantum sensors with existing medical imaging techniques, healthcare providers can gain more precise and reliable data to inform their decisions.

8.6 STANDARDS AND REGULATION ISSUES FOR QUANTUM SENSORS

The advent of quantum technologies has heralded a new era of scientific advancement, offering unprecedented capabilities in various fields, including sensing, computing, and communication. Quantum sensors, in particular, have demonstrated remarkable sensitivity and precision in measuring physical quantities such as magnetic fields, temperature, and time. However, as these technologies move closer to real-world applications, there is an increasing need for clear and effective standards and regulations to ensure their reliability, safety, and interoperability.

The rapid development of quantum sensor technologies poses unique challenges for regulatory bodies and standards organizations, requiring new frameworks that address the specificities of quantum systems while also maintaining compatibility with existing industrial standards. The development of standards and regulations for quantum sensors is essential for ensuring their safe, reliable, and ethical use in practical applications. As quantum sensing technologies continue to evolve, it is critical that standards organizations and regulatory bodies address the unique challenges posed by these systems. By establishing clear guidelines for safety, calibration, interoperability, and ethical considerations, stakeholders can help unlock the full potential of quantum sensors while mitigating risks and ensuring their responsible use. As the field of quantum sensing grows, ongoing collaboration between researchers, industry leaders, and policymakers will be key to developing the standards and regulations necessary for the successful integration of quantum sensors into real-world systems.

8.6.1 Challenges in Defining Standards for Quantum Sensors

One of the primary challenges in establishing standards for quantum sensors is the inherent complexity of quantum systems. Unlike classical systems, which follow predictable, deterministic behavior, quantum systems are governed by probabilistic principles, such as superposition and entanglement. These unique properties present difficulties in defining standardized measurement techniques and calibration methods. In many cases, quantum sensors function at the edge of current technological capabilities, making it challenging to develop universally applicable standards that can cover the full range of quantum sensor types, from those based on cold atoms to those utilizing superconducting qubits. Another challenge arises from the rapidly evolving nature of quantum technologies. The field of quantum sensing is still in its infancy, with new advancements and novel sensor designs emerging on a regular basis. As a result, standards for quantum sensors need to be flexible enough to accommodate these rapid advancements. This dynamic nature of the field requires standards and regulatory frameworks to be regularly updated to reflect the latest technological developments and best practices. Creating adaptable and forward-thinking standards is essential to ensure that quantum sensors can be effectively integrated into existing systems and used in a variety of real-world applications.

8.6.2 Safety and Reliability Concerns

As quantum sensors begin to be deployed in practical applications, safety and reliability become critical concerns. Quantum sensors often operate in highly sensitive environments, such as space exploration, medical diagnostics, and autonomous systems. The precision and sensitivity of these sensors make them vulnerable to environmental factors such as electromagnetic interference, temperature fluctuations, and vibrations. Ensuring that quantum sensors perform reliably under these conditions requires rigorous testing, calibration, and validation protocols, which should be outlined in formal standards. Additionally, because quantum sensors are based on quantum mechanics, they may exhibit unpredictable behavior in certain situations, such as during decoherence or entanglement loss. These factors could impact the accuracy of measurements and the overall performance of quantum sensor systems. Regulatory bodies need to address these challenges by setting clear guidelines for sensor calibration, error correction mechanisms, and performance testing under various conditions. These standards will ensure that quantum sensors meet the required performance criteria before they are deployed in sensitive or critical environments.

8.6.3 Interoperability with Classical Systems

Another key issue in the standardization of quantum sensors is ensuring interoperability with existing classical systems. While quantum sensors offer significant advantages in terms of sensitivity and accuracy, many practical applications still rely on classical technologies. For example, navigation systems often use classical gyroscopes and accelerometers, while medical imaging systems depend on classical sensors like MRI machines.

As quantum sensors begin to be integrated into these classical systems, it is essential to establish standards that ensure seamless communication and compatibility between quantum and classical technologies. Standards related to data formats, communication protocols, and interface design will be essential to facilitate the integration of quantum sensors into existing systems. Furthermore, quantum sensors often operate under different environmental conditions and require specific control systems to function properly. Standardizing these interfaces will ensure that quantum sensors can be easily integrated into real-world applications without requiring extensive modifications to classical systems. This interoperability will be crucial for realizing the full potential of quantum sensors in fields such as autonomous vehicles, environmental monitoring, and healthcare.

8.6.4 Ethical and Privacy Considerations

As quantum sensors become more widespread, there will also be ethical and privacy concerns related to their use. Quantum sensors, particularly those based on advanced imaging or sensing technologies, may enable the collection of highly detailed data about individuals or environments.

For instance, quantum sensors can be used for non-invasive medical diagnostics, but they may also raise concerns about the potential for unauthorized surveillance or data misuse. Establishing ethical guidelines and regulations for the use of quantum sensors is crucial to ensure that their benefits are realized without infringing on privacy rights or ethical boundaries. Moreover, the ability of quantum sensors to measure physical quantities with extreme precision could lead to unintended consequences in certain applications. For example, the use of quantum sensors in defense or security contexts could potentially enhance surveillance capabilities, raising concerns about their impact on civil liberties. Developing regulatory frameworks that balance the benefits of quantum sensors with the protection of individual rights will be an important challenge for policymakers and regulatory bodies.

CHAPTER 9

Quantum Sensing in Industry and Research

9.1 QUANTUM SENSORS FOR INDUSTRIAL APPLICATIONS

9.1.1 Semiconductor and Manufacturing Quality Control

The semiconductor industry is the backbone of modern electronics, with applications spanning across computing, communications, automotive, healthcare, and a variety of other sectors. As demand for increasingly powerful and energy-efficient devices grows, the importance of semiconductor manufacturing quality control (QC) has never been more significant. Quality control in semiconductor manufacturing ensures that products meet the necessary standards for performance, reliability, and safety. The production of semiconductors is a highly complex process that involves various intricate steps, from raw material processing to device fabrication. Given the precision required in semiconductor devices, any deviation from quality control protocols can lead to significant failures, defects, or performance issues, which could be costly both in terms of financial losses and reputation damage. Semiconductor manufacturing quality control is a critical component of ensuring the performance, reliability, and safety of semiconductor devices.

As the industry evolves, manufacturers must adapt their quality control processes to address the challenges presented by shrinking device sizes, contamination risks, and material variability. Automation, AI, and advanced measurement technologies will play an increasingly important role in improving the accuracy and efficiency of quality control systems. With the continued growth of the semiconductor industry, maintaining high standards of quality control will be essential for meeting the demands of emerging technologies and ensuring the continued success of the sector.

9.1.2 Smart Grid and Power Systems Monitoring

The modern power grid has evolved significantly over the past century, with technological advancements improving its efficiency, reliability, and ability to accommodate renewable energy sources. The traditional electrical grid is being transformed into a "smart grid," a system that incorporates digital communication technology to manage electricity more efficiently. This transformation is a crucial step toward ensuring a sustainable and resilient energy future, as it helps address challenges such as aging infrastructure, increased demand, and the integration of renewable energy sources like solar and wind power. At the heart of the smart grid lies the ability to monitor and control power systems in real-time. This monitoring involves the use of advanced sensors, communication networks, and data analytics to ensure the grid operates at optimal levels. Smart grid and power systems monitoring enable utilities to collect and analyze data in real-time, giving operators valuable insights into the functioning of the grid. By using advanced monitoring systems, operators can identify potential problems before they cause widespread disruptions, optimize energy flow, and improve the integration of distributed energy resources (DERs). The combination of intelligent sensors, automated control systems, and communication technologies allows the smart grid to self-regulate, adapt to changing conditions, and ultimately enhance the quality of service for consumers. The integration of smart grid and power systems monitoring technologies is revolutionizing the way electricity is generated, distributed, and consumed. By enabling real-time monitoring, advanced data analytics, and the integration of renewable energy sources, smart grids are making power systems more efficient, reliable, and resilient. As the world continues to transition to a more sustainable energy future, the role of smart grids will only grow in importance. The ability to monitor, control, and optimize energy flow will be crucial in meeting the demands of a rapidly changing energy landscape, ensuring that electricity is delivered where and when it is needed most.

9.2 QUANTUM SENSORS IN METROLOGY AND CALIBRATION LABS

Metrology, the science of measurement, plays a critical role in various industries, including manufacturing, healthcare, and research. Accurate and precise measurements are essential for ensuring the functionality and quality of products and systems. Traditional methods of measurement have been advanced over time, but the advent of quantum technology has brought about a revolution in the field of metrology. Quantum sensors, leveraging the principles of quantum mechanics, offer unprecedented levels of precision and sensitivity, making them a powerful tool in metrology and calibration laboratories. These sensors use quantum properties such as superposition, entanglement, and quantum coherence to measure physical quantities with exceptional accuracy, surpassing the limitations of classical sensors. The application of quantum sensors in metrology and calibration labs is particularly important in achieving more accurate standards of measurement, improving measurement reliability, and minimizing uncertainties in measurements. By harnessing the unique features of quantum states, these sensors can achieve precision that is beyond the capabilities of conventional technologies.

This capability has profound implications in a range of fields, from fundamental physics and quantum research to industrial applications such as manufacturing and healthcare diagnostics. Quantum sensors are set to revolutionize the field of metrology by offering unprecedented precision and sensitivity for a wide range of measurements. Their integration into calibration labs will enhance the accuracy of measurement standards, improve the reliability of measurement systems, and enable new scientific discoveries. Despite the challenges involved in their implementation, the continued development of quantum technologies will undoubtedly lead to new breakthroughs in metrology and calibration practices.

9.2.1 The Role of Quantum Sensors in Metrology

The integration of quantum sensors into metrology labs allows for more precise and reliable measurement of a variety of physical quantities. These sensors operate based on quantum mechanical phenomena, which provide an inherent advantage in terms of sensitivity and accuracy over traditional classical sensors. For instance, atomic clocks, which are based on quantum transitions of atoms, are the most accurate time-keeping devices available. They enable the definition of the international unit of time (second), a crucial standard for global measurements. Quantum sensors can be employed for a wide array of measurements, including magnetic fields, gravitational forces, temperature, and more. They rely on quantum mechanical effects, such as quantum interference and tunneling, to measure physical properties with a level of precision that was once thought unattainable. For example, quantum magnetometers exploit the properties of atomic spins to detect extremely weak magnetic fields, which are essential for applications in materials science, biomedical imaging, and geological exploration. The sensitivity of these quantum devices allows them to measure quantities that would be undetectable with classical techniques, making them invaluable for precision metrology. Another example is the use of quantum sensors in the measurement of acceleration and force. Devices like quantum accelerometers and gyroscopes use the principles of superposition and interference of quantum states to detect small changes in motion, which is fundamental for high-precision navigation systems and calibration of inertial sensors.

9.2.2 Advantages of Quantum Sensors in Calibration Labs

Calibration labs are tasked with ensuring that instruments and systems used for measurements adhere to defined standards of accuracy. The use of quantum sensors in these labs offers several distinct advantages over traditional methods. The most significant of these is the ability to achieve levels of precision that were previously unattainable. Quantum sensors can measure physical quantities with very low uncertainty, enabling them to provide more accurate calibration for a wide range of instruments. Traditional sensors often have limitations when it comes to sensitivity, especially in the measurement of small or weak signals. Quantum sensors, by leveraging quantum properties such as coherence and entanglement, can detect minute changes in physical properties, allowing for highly precise calibration. This capability is especially useful when calibrating instruments that are used for measuring quantities such as magnetic fields, time, and temperature, where even the smallest error can have significant consequences in terms of measurement accuracy. Furthermore, quantum sensors can be employed to establish more precise reference standards for various physical quantities. For instance, quantum-based atomic clocks are used as the gold standard for time measurement. By using these highly accurate clocks, calibration labs can provide more reliable time-based measurements and calibrate other time-sensitive instruments with greater precision.

9.2.3 Applications of Quantum Sensors in Calibration Labs

One of the primary applications of quantum sensors in calibration labs is in the calibration of time and frequency standards. Atomic clocks, based on quantum transitions in atoms such as cesium or rubidium, are the backbone of global timekeeping systems. These clocks provide an exceptionally stable and accurate reference for time measurements. Calibration labs use quantum sensors to compare and calibrate other timekeeping instruments, ensuring that all devices are aligned with the most precise and reliable time standards available. Quantum sensors also enable laboratories to generate more accurate frequency references, improving the calibration of devices such as oscillators, frequency counters, and communication systems. Another significant application is in the measurement of magnetic fields. Quantum magnetometers, which exploit the spin properties of atoms such as nitrogen-vacancy (NV) centers in diamonds or rubidium vapor, are used to measure extremely weak magnetic fields with high sensitivity.

These magnetometers are essential for calibrating instruments used in applications such as magnetic resonance imaging (MRI), geological surveys, and materials characterization. Calibration labs use quantum magnetometers to establish more accurate magnetic field standards and ensure that magnetic sensors and devices are functioning correctly. In the field of force and acceleration measurement, quantum accelerometers are increasingly being used in calibration labs. These devices rely on the principles of quantum interference and superposition to detect small changes in acceleration and force. By using quantum sensors to calibrate accelerometers and inertial navigation systems, labs can improve the precision of devices used in high-stakes applications such as aerospace navigation, seismic sensing, and geophysical exploration. Quantum accelerometers offer an unprecedented level of accuracy and sensitivity, making them essential for precision measurement in industries that rely on highly accurate motion sensing.

9.2.4 Challenges and Considerations in Quantum Sensor Implementation

While quantum sensors offer remarkable precision and accuracy, their implementation in calibration labs is not without challenges. One of the key obstacles is the need for highly controlled environments to maintain quantum coherence. Quantum systems are sensitive to environmental disturbances, such as temperature fluctuations, electromagnetic interference, and mechanical vibrations, which can lead to decoherence and loss of sensitivity. For quantum sensors to function effectively in metrology and calibration labs, specialized conditions must be maintained, including shielding from external noise and vibration isolation. As such, there is an ongoing need for research into more robust and resilient quantum sensor designs that can operate in a broader range of environments. Additionally, quantum sensors often require sophisticated and costly equipment for their operation, which can present financial and logistical challenges for calibration labs. The need for cryogenic temperatures, vacuum chambers, or high-precision lasers to manipulate quantum states adds complexity to the implementation of quantum sensors. Overcoming these technical and economic barriers will be crucial for the widespread adoption of quantum sensors in practical metrology applications.

9.3 QUANTUM SENSING FOR DEFENSE AND SECURITY

Quantum sensing is an emerging field that harnesses the principles of quantum mechanics to measure physical quantities with unprecedented precision and sensitivity. In the domains of defense and security, the applications of quantum sensing are poised to bring about revolutionary advances, allowing for more accurate and efficient monitoring of critical systems and environments. Unlike classical sensors, which are limited by conventional measurement techniques, quantum sensors take advantage of quantum phenomena such as superposition, entanglement, and quantum coherence to offer capabilities that far surpass their classical counterparts. These advancements open new possibilities for a range of defense and security applications, from detecting hidden threats to enhancing communication and navigation systems.

The integration of quantum sensors into defense and security systems holds immense promise. With their superior sensitivity and accuracy, quantum sensors can provide highly reliable measurements in complex and dynamic environments, making them indispensable in modern defense applications. From precision navigation systems to secure communication and enhanced threat detection, quantum sensing is on the verge of transforming the way military and security operations are conducted. However, with these advancements come challenges related to the deployment, scaling, and integration of quantum technologies into existing systems. Quantum sensing holds immense promise for transforming defense and security applications. From enhancing threat detection and secure communication to enabling GPS-free navigation and improving surveillance capabilities, quantum sensors offer unprecedented levels of precision and sensitivity. While there are still technical challenges to overcome, the potential benefits of quantum sensing in defense and security are vast, and their integration into operational systems is likely to shape the future of military and security operations worldwide.

9.3.1 Quantum Sensors for Threat Detection and Surveillance

One of the most significant applications of quantum sensing in defense and security is in the area of threat detection and surveillance. Classical detection technologies, such as radar and sonar, rely on conventional signals and often suffer from limitations in sensitivity and resolution, particularly in detecting low-signature or stealth objects. Quantum sensors, on the other hand, can detect faint signals with exceptional precision, making them invaluable in detecting threats that would otherwise go unnoticed. Quantum-enhanced magnetometers are one example of how quantum sensors are used for threat detection. These sensors can measure tiny magnetic fields with extremely high sensitivity, enabling the detection of submarines, concealed weapons, or other military assets that generate weak magnetic signals. Unlike classical magnetic sensors, which are limited by noise and interference, quantum magnetometers can detect these signals in challenging environments, such as underwater or within heavily shielded areas. This capability is particularly valuable in defense operations, where detecting hidden or moving threats in real-time is crucial for operational success.

Another important application of quantum sensors in threat detection involves the use of quantum-enhanced accelerometers and gyroscopes for precise navigation and motion tracking. In military applications, it is essential to accurately track the movement of vehicles, drones, and other assets in complex environments, especially when GPS signals are unavailable or jammed. Quantum accelerometers and gyroscopes, such as those based on cold atom interferometry, offer extreme sensitivity in measuring small changes in velocity and orientation. These sensors can operate in GPS-denied environments, providing highly accurate navigation and positioning data even in the most challenging conditions.

9.3.2 Quantum Communication and Secure Transmission

In the realm of defense and security, the protection of sensitive information is of utmost importance. Traditional communication systems, while effective, are vulnerable to a variety of security threats, including hacking, eavesdropping, and interception. Quantum communication technologies offer a promising solution to these challenges by utilizing the principles of quantum mechanics to secure data transmission in ways that classical systems cannot match. Quantum key distribution

(QKD) is one of the most widely studied quantum communication techniques. QKD relies on the principles of quantum entanglement and the no-cloning theorem to ensure that any attempt to intercept or eavesdrop on the communication will be detectable. In QKD, the sender and receiver share a cryptographic key using quantum entangled particles, such as photons. If an eavesdropper tries to intercept the key, the quantum state of the particles will be altered, revealing the presence of the intrusion. This property ensures the security of the transmitted data and protects sensitive military and governmental communications from unauthorized access. Quantum communication can also enable secure communication networks that are immune to traditional hacking methods. By employing quantum entanglement and teleportation, quantum communication systems can achieve levels of security that are practically impossible to break with classical computational techniques. These advancements have significant implications for defense, where secure communication between military units, command centers, and intelligence agencies is essential for effective coordination and operational success.

9.3.3 Quantum Sensing for GPS-Free Navigation

In military and defense operations, precise navigation is critical, particularly in environments where GPS signals are weak or denied, such as in urban canyons, underground, or in hostile territories. Traditional navigation systems rely heavily on GPS satellites, but GPS jamming and spoofing are common security concerns. Quantum sensors offer a promising alternative by providing highly accurate and reliable positioning and navigation data without the need for GPS. Quantum-enhanced inertial sensors, such as quantum gyroscopes and accelerometers, can be used to create navigation systems that do not rely on external signals.

These sensors are based on quantum interference effects, which enable them to measure small changes in motion with extreme precision. For example, quantum gyroscopes use the interference of quantum states to detect rotational movements, providing accurate heading and orientation data even in environments where GPS signals are unavailable or unreliable. The development of quantum-based inertial navigation systems could enable autonomous vehicles, drones, and military aircraft to navigate with extreme accuracy without the need for GPS. This capability is especially important in defense applications, where GPS signals can be intentionally disrupted or blocked by adversaries. Quantum sensors, by providing a GPS-free navigation solution, would allow military assets to operate with greater independence and security in challenging environments.

9.3.4 Quantum Sensors in Surveillance and Reconnaissance

Quantum sensors also have significant potential in surveillance and reconnaissance missions, where the ability to monitor and gather intelligence from the field is essential. Quantum-enhanced imaging systems, which utilize quantum properties such as entanglement and superposition, can provide higher resolution and greater sensitivity than conventional imaging systems. For example, quantum-enhanced radar and lidar systems can be used for remote sensing and imaging of objects in difficult-to-reach areas. These quantum sensors take advantage of quantum interference and other quantum phenomena to detect small objects, measure distances with greater precision, and even detect hidden objects in environments that are otherwise challenging for classical sensors to penetrate. These advanced imaging systems are invaluable for military reconnaissance, border security, and surveillance operations, where detecting hidden threats or identifying targets in complex environments is crucial. In addition, quantum sensors can be employed for environmental monitoring in defense applications. For instance, quantum sensors can be used to detect chemical, biological, or radiological threats in the air or on the ground. By providing early warning of potential hazards, quantum sensors help ensure the safety of military personnel and civilians in conflict zones or hazardous environments.

9.4 THE ROLE OF QUANTUM SENSORS IN SPACE EXPLORATION

9.4.1 Space-Based Quantum Sensors for Earth Observation

Space-based quantum sensors represent a cutting-edge development in the field of remote sensing and Earth observation. These sensors leverage the principles of quantum mechanics to offer exceptional precision and sensitivity in measuring physical phenomena such as gravity, magnetic fields, and electromagnetic waves. Traditional Earth observation technologies, such as radar and optical imaging systems, are limited by factors like resolution, noise, and interference. In contrast, quantum sensors exploit quantum properties such as superposition, entanglement, and quantum coherence, allowing for unprecedented accuracy in measurements.

The potential applications of space-based quantum sensors for Earth observation span a wide range of fields, including climate monitoring, environmental surveillance, resource management, and disaster prediction. By harnessing quantum technologies, space-based sensors can offer new insights into Earth's dynamic processes and help address some of the most pressing challenges of the 21st century.

The integration of quantum sensors into space-based platforms promises to revolutionize how we monitor and understand our planet. These sensors can provide data with unmatched precision, improving our ability to observe and respond to environmental changes, track natural resources, and predict hazardous events. However, there are several challenges that must be addressed, including the complexity of deploying quantum sensors in space and their susceptibility to external interference. Despite these obstacles, the potential of space-based quantum sensors is immense, and they are poised to play a pivotal role in future Earth observation missions.

9.4.2 Quantum Sensors for Space Missions

Quantum sensors have the potential to transform space exploration and missions by offering unprecedented precision and sensitivity. These sensors take advantage of quantum mechanical properties such as superposition, entanglement, and quantum interference to achieve high levels of measurement accuracy beyond what classical sensors can provide. As space missions grow more ambitious, from lunar exploration to deep-space probes, quantum sensors promise to play a crucial role in various mission objectives, including navigation, Earth observation, and astrophysical studies. Their capability to measure physical phenomena such as gravity, magnetic fields, and time with extreme precision opens up a new frontier for space-based science. The integration of quantum sensors in space missions could vastly improve our understanding of the universe, enhance mission success rates, and enable new scientific discoveries that were previously unattainable with traditional sensors.

CHAPTER 10

Emerging Trends in Quantum Sensing

10.1 QUANTUM SENSING FOR AUTONOMOUS SYSTEMS AND VEHICLES

Quantum sensing represents a significant technological leap in enhancing the capabilities of autonomous systems and vehicles. As autonomous systems—ranging from self-driving cars to unmanned aerial vehicles (UAVs) and underwater drones—become more integral to modern industries and daily life, the demand for greater precision, reliability, and resilience in their navigation and environment-sensing tools continues to grow.

Traditional sensors such as GPS, LiDAR, radar, and inertial measurement units (IMUs) form the backbone of current autonomous systems. However, these technologies have limitations, particularly in environments with poor signal availability or adverse conditions. Quantum sensors, with their ability to measure fundamental physical quantities such as acceleration, rotation, magnetic fields, and gravity at unprecedented sensitivities, offer an opportunity to overcome many of these constraints and redefine the landscape of autonomous navigation and situational awareness.

10.1.1 Principles of Quantum Sensing Relevant to Mobility

Quantum sensing relies on exploiting the principles of quantum mechanics, such as superposition, entanglement, and quantum coherence, to achieve highly accurate measurements. In autonomous systems, the most applicable quantum sensors include quantum accelerometers, gyroscopes, magnetometers, and gravimeters. These instruments can offer precise data about motion and orientation without reliance on external signals. For example, a quantum accelerometer measures the acceleration of an object with extraordinary accuracy by observing the interference patterns of matter waves, typically using ultra-cold atoms. Similarly, quantum gyroscopes provide superior rotational sensing by tracking phase shifts in quantum states caused by angular motion. These types of sensors are especially beneficial in situations where GPS is unavailable or unreliable, such as inside tunnels, underwater, or in urban environments with signal obstruction.

10.1.2 Replacing and Complementing Traditional Navigation Systems

One of the most prominent applications of quantum sensors in autonomous vehicles lies in navigation systems. Currently, autonomous systems rely heavily on satellite-based positioning systems like GPS. However, these signals can be disrupted by urban canyons, jamming, spoofing, or natural obstructions, leading to significant operational risks. Quantum inertial navigation systems (QINS) can mitigate this vulnerability. By using quantum accelerometers and gyroscopes, QINS offer self-contained navigation solutions that do not require external references. This means that vehicles equipped with QINS can maintain accurate location and trajectory information over long durations, even in GPS-denied environments. Quantum sensors can also enhance traditional inertial navigation systems by reducing drift, a common issue with classical accelerometers and gyroscopes. Classical sensors accumulate error over time, leading to significant positional inaccuracies. Quantum sensors, on the other hand, exhibit lower noise and longer coherence times, enabling more stable and drift-resistant navigation data. This synergy between classical and quantum sensors creates robust hybrid navigation systems that offer both redundancy and improved precision.

10.1.3 Enhancing Autonomous Vehicle Perception and Environmental Awareness

In addition to navigation, quantum sensors have the potential to revolutionize how autonomous vehicles perceive and interpret their surroundings. Quantum magnetometers and gravimeters can detect subtle changes in magnetic fields or gravitational gradients, enabling the creation of highly detailed and localized maps of the environment. These measurements can be used to enhance SLAM (Simultaneous Localization and Mapping) techniques, a core component of autonomous vehicle operation. For instance, a quantum gravimeter can detect underground structures or variations in terrain that are invisible to traditional sensors.

This information could be crucial for autonomous vehicles operating in challenging environments such as mining operations, geological surveys, or military reconnaissance missions. Quantum magnetometers can similarly provide data about nearby metallic objects or infrastructure, contributing to obstacle avoidance and navigation in complex environments. The use of quantum sensing in these domains allows autonomous systems to operate with a higher level of environmental intelligence and adaptability.

10.1.4 Applications in Air, Sea, and Space Autonomous Vehicles

The applicability of quantum sensing spans various domains of autonomous vehicle operation, including air, sea, and space. For aerial drones, especially those used in defense or surveillance, quantum gyroscopes and accelerometers provide navigation capabilities without depending on GPS, which can be vulnerable to jamming. This resilience is critical for military and high-security applications. In addition, quantum magnetometers can help UAVs detect underground or hidden threats based on magnetic anomalies, enhancing reconnaissance missions. Underwater autonomous vehicles face a unique set of challenges, such as the inability to use GPS and limited visibility. Quantum sensors can address these limitations by enabling precise inertial navigation and environmental sensing based on gravitational or magnetic field variations.

For example, a quantum gravimeter could help a submersible detect underwater topographical features or navigate based on subtle differences in gravitational pull. In space, autonomous probes and rovers benefit from quantum sensors for navigation and positioning in regions where traditional systems are impractical. On planets or moons without satellite infrastructure, quantum inertial navigation could enable precise localization and movement. Furthermore, quantum sensors could detect gravitational anomalies or magnetic field variations on extraterrestrial surfaces, aiding in planetary exploration and scientific research.

10.1.5 Integration Challenges and Technological Hurdles

Despite the immense potential of quantum sensors, their integration into autonomous systems poses several technological challenges. One of the most significant hurdles is miniaturization.

Quantum sensors often require complex setups involving lasers, vacuum systems, and cryogenics, which are difficult to compact into the limited space available in autonomous vehicles. Advances in photonics, microfabrication, and chip-scale atomic devices are gradually addressing these issues, but widespread deployment will depend on further engineering innovations. Environmental robustness is another concern. Autonomous vehicles operate in varied and sometimes harsh conditions, including extreme temperatures, vibrations, and electromagnetic interference. Quantum sensors must be engineered to maintain their precision and stability under these operational stresses. Developing ruggedized quantum sensors that can withstand real-world conditions while retaining their quantum coherence is essential for their broader adoption.

Furthermore, integrating quantum sensors into existing control architectures of autonomous systems requires compatible software and data processing frameworks. Quantum sensors generate different types of data with unique noise characteristics and error models compared to classical sensors. This necessitates the development of new algorithms and sensor fusion strategies to interpret and utilize quantum data effectively within autonomous decision-making systems.

10.2 THE FUTURE OF QUANTUM IMAGING AND REMOTE SENSING

Quantum imaging and remote sensing are rapidly emerging fields that leverage the principles of quantum mechanics to revolutionize how we capture, transmit, and interpret information about distant or concealed environments. Unlike classical imaging and sensing technologies that depend on macroscopic interactions of light and matter, quantum techniques utilize quantum phenomena such as entanglement, superposition, and quantum interference to surpass conventional limits of resolution, sensitivity, and information content. As advancements in quantum technologies continue to unfold, the potential for quantum imaging and remote sensing to impact diverse domains—including environmental monitoring, defense, space exploration, and biomedical diagnostics—is becoming increasingly evident. The future of these technologies promises not only to enhance current capabilities but to unlock entirely new modalities of perception and measurement.

10.2.1 Fundamentals of Quantum Imaging

Quantum imaging encompasses a range of techniques that exploit the quantum properties of light to enhance image formation. One of the most notable approaches is quantum ghost imaging, in which an image is reconstructed from correlated photon pairs, even though the detector collecting spatial information never directly sees the object. This technique challenges classical intuitions and offers the ability to form images in conditions where conventional cameras would fail, such as low-light or highly scattered environments. Another important concept is quantum illumination, where entangled or correlated photon pairs are used to detect the presence of objects with enhanced signal-to-noise ratio, particularly in noisy or lossy environments. These principles demonstrate how quantum resources can be harnessed to improve imaging contrast, resolution, and resilience. As quantum light sources and detectors become more efficient and practical, quantum imaging will likely become an integral part of next-generation sensing systems. Single-photon detectors, squeezed light sources, and photon-number-resolving technologies are all critical components that enable quantum imaging beyond classical constraints. Furthermore, advancements in quantum optics and integrated photonics are driving the miniaturization and robustness of quantum imaging systems, making them more viable for deployment in real-world applications.

10.2.2 Applications in Environmental and Earth Observation

One of the most promising areas for the application of quantum imaging and remote sensing is environmental monitoring and Earth observation. Satellites and airborne systems equipped with quantum-enhanced sensors could provide unprecedented detail in mapping atmospheric pollutants, detecting changes in vegetation, monitoring water quality, and identifying signs of climate change.

Quantum sensors have the ability to detect minute spectral signatures and weak signals with high accuracy, making them suitable for measuring trace gases or subtle variations in the Earth's magnetic or gravitational fields. In addition to monitoring large-scale environmental changes, quantum imaging techniques can be applied to localized inspections of ecosystems, such as tracking wildlife movement through canopy layers or assessing soil health through spectral imaging. Quantum-enhanced LiDAR systems, which utilize entangled photons or squeezed light to improve depth resolution and range accuracy, offer a significant improvement over traditional LiDAR technologies, especially in foggy or low-light conditions. These advancements could greatly benefit agriculture, forestry, and disaster management by enabling more precise and timely data collection.

10.2.3 Quantum Imaging in Defense and Security

In the field of defense and national security, quantum imaging offers capabilities that are difficult or impossible to achieve with classical methods. Quantum illumination and ghost imaging can be used for stealthy surveillance and target detection, as they allow the detection of objects in highly cluttered environments with reduced probability of interception. For instance, quantum radar systems can detect objects with low reflectivity or camouflage by using entangled photons to distinguish between signal and noise, even when conventional radar would fail. Furthermore, quantum imaging can be applied to see through partially opaque or obscured environments, such as smoke, fog, or foliage.

This ability can be critical during rescue missions, battlefield operations, or covert reconnaissance. Quantum-enhanced thermal imaging, using quantum correlations to improve temperature resolution, also holds potential for identifying hidden heat sources or human activity. These technologies could redefine the capabilities of future security systems by enabling precise, low-light, and non-invasive imaging across a variety of scenarios.

10.2.4 Biomedical Imaging and Diagnostics

The future of medical imaging and diagnostics is also likely to benefit greatly from quantum technologies. Quantum imaging allows for reduced exposure to harmful radiation while still achieving high-resolution images. Techniques like quantum-enhanced optical coherence tomography (OCT) and quantum fluorescence imaging could provide deeper insights into biological tissues, allowing early detection of diseases such as cancer or neurological disorders. The use of entangled photons can improve image clarity in biological samples by reducing noise and enabling better contrast in structures at the cellular and molecular levels. Moreover, the potential for quantum sensors to detect extremely weak electromagnetic fields could lead to breakthroughs in functional brain imaging or cardiac diagnostics. These advances would not only improve patient outcomes through earlier diagnosis but also enable real-time, non-invasive monitoring of physiological processes. The integration of quantum imaging into medical instruments may ultimately lead to safer, more accurate, and more accessible healthcare technologies.

10.2.5 Space Exploration and Astronomy

Quantum imaging and sensing are poised to play a pivotal role in space-based observations and astronomical research. Space missions require instruments capable of detecting extremely weak signals from distant celestial bodies or cosmic phenomena. Quantum technologies, particularly those involving single-photon detection and quantum entanglement, allow for the collection of high-fidelity data from the far reaches of space. Quantum interferometry, for instance, can be used to construct large baseline telescopes that offer unprecedented angular resolution, enabling the study of exoplanets, black holes, and cosmic microwave background radiation with greater precision. Additionally, quantum remote sensing satellites can be used for Earth observation from orbit, providing real-time monitoring of environmental and geological phenomena. The long-range transmission of quantum-entangled photons between Earth and satellites also opens the door for secure quantum communication and teleportation experiments in space. As quantum payloads become more compact and robust, they are increasingly suitable for space missions, where size, weight, and power constraints are paramount.

10.3 QUANTUM SENSING IN AI AND MACHINE LEARNING APPLICATIONS

Quantum sensing, a frontier technology rooted in the principles of quantum mechanics, has seen increasing relevance in the context of artificial intelligence (AI) and machine learning (ML). These quantum sensors exploit phenomena such as entanglement, superposition, and quantum tunneling to measure physical quantities with extraordinary precision and sensitivity. When combined with AI and ML, quantum sensing can offer novel capabilities in data acquisition, analysis, and decision-making. This integration represents a symbiotic relationship where quantum sensors provide high-fidelity data inputs, and AI algorithms interpret, optimize, and adapt sensor functionalities in real-time. As industries and research fields continue to rely on intelligent systems, the convergence of quantum sensing with AI and ML could lead to groundbreaking innovations across sectors such as healthcare, autonomous systems, environmental monitoring, and advanced manufacturing.

10.3.1 Quantum Sensors as Data Sources for AI Systems

AI and ML systems thrive on data quality and volume, and quantum sensors are uniquely positioned to deliver both. Traditional sensors often struggle with noise, resolution limits, and environmental interference, but quantum sensors can detect subtle variations in electromagnetic fields, temperature, gravitational forces, and time intervals with unprecedented precision. These capabilities are especially valuable in scenarios requiring high-resolution datasets, such as brain-computer interfaces, real-time environmental mapping, and structural health monitoring. The ability of quantum sensors to detect minute anomalies or changes enhances the predictive power of AI models, which depend on accurate and granular inputs for effective training and real-world performance. By using quantum sensors to gather data, AI algorithms can be trained on more reliable and extensive datasets. For instance, in the field of remote sensing, quantum magnetometers and gravimeters can offer more detailed terrain and resource data. When fed into ML models, this information can enable advanced pattern recognition, resource forecasting, and anomaly detection. Quantum-enhanced data inputs can also improve reinforcement learning environments, especially in simulations used for autonomous navigation, climate modeling, or financial market analysis.

10.3.2 Machine Learning for Optimizing Quantum Sensor Performance

While quantum sensors are inherently sensitive and accurate, they are also susceptible to noise, drift, and operational instability, particularly in non-laboratory settings. This is where machine learning plays a vital role. AI techniques, especially supervised and unsupervised learning, can be used to calibrate, stabilize, and correct quantum sensor outputs in real-time. For example, ML models can learn to distinguish between meaningful signal variations and noise introduced by environmental factors or system imperfections. By continuously analyzing sensor outputs and comparing them with expected baselines, these models can adjust operational parameters, apply correction algorithms, and even predict when recalibration is needed. Adaptive machine learning models, such as neural networks and Gaussian processes, can dynamically learn from the performance history of quantum sensors. This enables predictive maintenance and real-time system tuning, ensuring that sensors function optimally even in fluctuating conditions. Additionally, AI can be used to automate the alignment, synchronization, and readout procedures for complex quantum sensing systems, thus reducing human error and making deployment easier in field operations. The incorporation of ML into sensor management creates a feedback loop that enhances both the reliability and longevity of quantum sensing instruments.

10.3.3 Quantum Sensing in Smart Infrastructure and IoT

The proliferation of smart infrastructure and Internet of Things (IoT) systems has created an ecosystem where billions of interconnected devices collect and exchange information. In this environment, quantum sensors can play a transformative role by offering enhanced accuracy and resilience to noise and environmental changes. For instance, quantum accelerometers and gyroscopes can improve navigation in GPS-denied environments, while quantum-based timekeeping can synchronize systems with nanosecond precision. When integrated with AI-driven IoT platforms, quantum sensors allow for smarter, more autonomous responses to environmental stimuli. AI algorithms can manage the vast data streams generated by quantum-enhanced IoT networks, identifying trends, forecasting issues, and making autonomous decisions. For example, in a smart city scenario, quantum sensors can monitor air quality or detect structural weaknesses in buildings. AI can analyze this information in real-time to alert authorities or automatically adjust public systems like ventilation, lighting, or traffic flow. The combination of quantum sensing and intelligent automation leads to more efficient, adaptive, and resilient infrastructures capable of evolving with changing urban needs.

10.3.4 Applications in Healthcare and Biomedical Fields

Quantum sensors are poised to revolutionize medical diagnostics and monitoring by providing extremely sensitive measurements of physiological signals. Magnetoencephalography (MEG) and magnetic resonance imaging (MRI) are examples where quantum sensors can significantly enhance imaging resolution and accuracy. When paired with AI algorithms trained on medical datasets, quantum sensors enable early detection of neurological disorders, cardiovascular anomalies, and other complex conditions that may be invisible to classical techniques. Machine learning models can analyze the high-resolution data from quantum sensors to identify subtle biomarkers and predict disease progression with greater confidence. For instance, AI can detect patterns in brain wave fluctuations measured by quantum magnetometers, potentially aiding in the diagnosis of epilepsy, Alzheimer's disease, or traumatic brain injury. Wearable devices embedded with quantum sensors could continuously monitor patients' vital signs, and AI could provide personalized health insights or early warnings. This integration opens the door to proactive healthcare, where real-time data analysis and precision sensing work together to optimize treatment and improve patient outcomes.

10.3.5 Enhancing Autonomy in Vehicles and Robotics

Autonomous vehicles and robotics demand precise perception and navigation capabilities, especially in complex and unpredictable environments. Quantum sensors can contribute by providing highly accurate inertial measurements, magnetic field detection, and environmental mapping. Quantum-enhanced inertial navigation systems can function effectively even without GPS, which is crucial for applications in underwater exploration, space missions, and military operations. AI algorithms can fuse quantum sensor data with classical sources like LiDAR, cameras, and radar, creating a multi-modal perception system that is more robust and intelligent. Reinforcement learning and deep learning models can process this integrated sensory input to make decisions, navigate obstacles, and execute complex tasks with greater efficiency. For example, a drone equipped with quantum accelerometers and magnetometers could map an unknown environment while AI algorithms build a real-time 3D model, identify objects of interest, and adjust flight paths dynamically. In industrial robotics, this combination could improve precision, reduce energy consumption, and enhance safety in human-robot collaboration scenarios.

10.3.6 Security, Defense, and Surveillance Applications

Quantum sensors are naturally suited for detecting subtle signals and changes, making them valuable tools for security and surveillance. Quantum gravimeters and magnetometers can detect underground structures, concealed weapons, or unauthorized movements with high sensitivity. AI and ML play a crucial role in interpreting this data, filtering out background noise, and identifying potential threats. In defense applications, AI-enhanced quantum radar systems can detect stealth aircraft or objects in noisy environments by analyzing correlations in entangled photon streams. Machine learning algorithms can also optimize the deployment of quantum sensors in surveillance networks, determining the best locations, operating parameters, and detection thresholds. In cyber-physical security scenarios, AI can correlate quantum sensor inputs with other security data, such as biometric verification or behavioral analysis, to enhance threat detection and response mechanisms. This fusion of quantum precision with AI intelligence enables smarter, faster, and more secure monitoring systems.

10.4 QUANTUM SENSORS IN HEALTHCARE AND PERSONALIZED MEDICINE

Quantum technologies, long considered the domain of physicists and high-energy laboratories, are now making significant strides into practical, real-world applications. Among these technologies, quantum sensors are emerging as powerful tools in healthcare and personalized medicine. These sensors leverage the unique properties of quantum mechanics, such as superposition and entanglement, to achieve sensitivities and precisions that classical sensors simply cannot match. This enhanced capability holds the potential to revolutionize diagnostics, monitoring, and treatment personalization, transforming how healthcare is delivered. Quantum sensors operate on the principles of quantum physics. Unlike classical sensors, which rely on conventional measurement techniques, quantum sensors exploit the quantum states of particles—such as atoms, photons, or electrons—to detect minute changes in the environment. These changes could include variations in magnetic fields, temperature, or other physical parameters that are often difficult to measure with traditional methods. Because quantum states are highly sensitive to external disturbances, these sensors can detect changes at the atomic or even subatomic scale, offering unprecedented precision. In the healthcare sector, this precision can be transformative. Detecting diseases in their earliest stages, monitoring subtle physiological changes, and accurately mapping biological structures at the molecular level are just a few applications that quantum sensors make possible. With their superior sensitivity and accuracy, quantum sensors provide new opportunities for non-invasive and real-time diagnostics, which are critical for the advancement of personalized medicine.

10.4.1 Early Disease Detection and Diagnosis

One of the most promising applications of quantum sensors in healthcare is in the early detection of diseases. Conditions such as cancer, neurodegenerative disorders, and cardiovascular diseases often remain undetected until they have progressed significantly, at which point treatment becomes more complex and less effective. Quantum sensors can be employed to detect the faintest biological markers associated with such diseases, enabling clinicians to diagnose them at a much earlier stage. For instance, quantum sensors based on nitrogen-vacancy centers in diamonds can detect extremely weak magnetic fields generated by neural or cardiac activity. These sensors can be used to monitor brain functions or detect abnormal heart rhythms with high spatial and temporal resolution. Furthermore, quantum-enhanced imaging techniques can allow for the visualization of cellular and molecular structures without damaging tissues, which is especially useful for studying sensitive organs like the brain. This can pave the way for earlier diagnoses of diseases such as Alzheimer's or Parkinson's by capturing changes in brain structure and function before clinical symptoms appear.

10.4.2 Advanced Imaging and Monitoring

Quantum sensors also enhance imaging modalities in medicine. Magnetic resonance imaging (MRI), one of the most vital tools in modern diagnostics, stands to benefit greatly from quantum technology. Quantum-enhanced MRI can improve the resolution and reduce the noise in imaging, offering more detailed and informative scans. This capability is crucial not just for identifying abnormalities, but also for monitoring how they evolve over time, especially in response to treatment. Additionally, quantum sensors can be used in magnetoencephalography (MEG) and magnetocardiography (MCG), which measure the magnetic fields generated by neuronal and cardiac activity, respectively.

Traditional sensors for these applications often require cryogenic cooling and are bulky. In contrast, quantum sensors can operate at room temperature and provide more accurate measurements, enabling compact, wearable, or portable diagnostic devices. These features make real-time, continuous health monitoring a feasible reality, even outside of hospital settings.

10.4.3 Enabling Personalized Medicine

Personalized medicine seeks to tailor treatment and prevention strategies to individual genetic, environmental, and lifestyle factors. To achieve this, highly detailed and individualized physiological data are required—something quantum sensors are particularly well-suited to provide. Their ability to detect subtle biochemical changes, track drug interactions at the molecular level, and monitor real-time physiological responses makes them invaluable tools in the personalization of healthcare. For example, quantum sensors can measure how different individuals metabolize drugs, providing critical information on appropriate dosages and reducing the risk of adverse drug reactions. Similarly, in cancer treatment, these sensors can help determine how a tumor responds to a specific therapy, allowing oncologists to adjust the treatment plan dynamically. This responsive approach enhances therapeutic efficacy and minimizes unnecessary side effects, thereby improving patient outcomes. Furthermore, wearable quantum devices could continuously monitor biomarkers, such as glucose levels in diabetic patients or inflammatory markers in autoimmune diseases. These devices can send real-time data to healthcare providers, who can then make informed decisions and adjust treatment regimens accordingly. Such integration of quantum sensing with digital health platforms could redefine chronic disease management and patient engagement.

10.4.4 Challenges and Research Directions

Despite their enormous potential, the implementation of quantum sensors in clinical settings is not without challenges. Quantum technologies are still relatively new and complex, often requiring sophisticated infrastructure and technical expertise. Scalability, reliability, and cost-effectiveness are major hurdles that researchers and engineers are working to overcome. Moreover, integrating quantum sensors into existing medical systems demands rigorous standardization, regulatory approval, and clinician training. Another significant challenge lies in data interpretation.

Quantum sensors generate highly precise and voluminous data, which must be analyzed accurately to derive meaningful clinical insights. This necessitates the development of advanced computational tools, such as machine learning algorithms and quantum computing interfaces, to process and interpret sensor data efficiently. Interdisciplinary collaboration between physicists, biomedical engineers, computer scientists, and clinicians is essential to translate quantum sensing technology from the lab to the bedside. On the research front, ongoing efforts are focused on miniaturizing quantum sensors, improving their sensitivity, and developing robust packaging for integration into medical devices. Initiatives are also underway to explore new materials and designs that enhance the stability and performance of quantum systems under physiological conditions. As the field progresses, it is anticipated that these developments will lead to more user-friendly, affordable, and widely deployable quantum healthcare technologies.

10.5 INTEGRATION OF QUANTUM SENSORS IN INTERNET OF THINGS (IOT) NETWORKS

The convergence of quantum technologies and the Internet of Things (IoT) marks a groundbreaking frontier in modern technological evolution. Among the most promising developments in this intersection is the integration of quantum sensors into IoT networks. Quantum sensors, leveraging the principles of quantum mechanics such as superposition, entanglement, and quantum tunneling, offer unparalleled sensitivity and precision in detecting physical quantities. These advanced capabilities significantly enhance the potential of IoT systems, enabling more accurate data collection, improved decision-making, and new applications across a broad range of sectors including healthcare, defense, smart cities, environmental monitoring, and industrial automation.

10.5.1 Understanding Quantum Sensors and IoT

Quantum sensors represent a new class of measurement devices that operate based on quantum phenomena. Unlike classical sensors, which depend on macroscopic properties and often face limitations in precision, quantum sensors utilize the intrinsic properties of quantum systems to detect extremely small variations in physical parameters like magnetic fields, gravitational forces, temperature, acceleration, and time. This exceptional sensitivity makes them ideal for high-precision applications that require detection of minute changes that conventional sensors might miss. On the other hand, the Internet of Things is a digital ecosystem of interconnected devices, sensors, and systems that communicate and exchange data over the internet. IoT networks rely heavily on the accurate sensing of environmental and operational conditions, feeding real-time data into analytics platforms and enabling automation, control, and optimization of various processes. Integrating quantum sensors into IoT networks thus represents a powerful enhancement, transforming the reliability and scope of IoT capabilities.

10.5.2 Enhancing IoT Capabilities with Quantum Sensing

The integration of quantum sensors into IoT systems enhances the accuracy, sensitivity, and reliability of data collected by these networks. In conventional IoT systems, the quality of data can be limited by sensor sensitivity and noise, leading to inaccuracies in system responses and decisions. Quantum sensors, by contrast, offer high fidelity measurements even in noisy environments, ensuring more dependable data acquisition. For instance, in precision agriculture, quantum sensors can detect subtle changes in soil composition, moisture levels, and atmospheric conditions, enabling farmers to make informed decisions about irrigation, fertilization, and crop protection. When embedded in an IoT framework, these sensors can provide continuous real-time data, enhancing productivity and sustainability. Similarly, in structural health monitoring, quantum accelerometers and gravimeters can detect microscopic shifts or stress points in bridges and buildings, allowing for predictive maintenance and disaster prevention. Furthermore, quantum-enhanced positioning and navigation systems integrated with IoT can achieve centimeter-level accuracy without relying on GPS, which is vulnerable to interference and spoofing. This has significant implications for autonomous vehicles, drones, and robotic applications, where precision and reliability are paramount. With quantum sensors providing highly accurate positioning data, IoT systems can operate more safely and efficiently.

10.5.3 Challenges of Integration and Interoperability

While the benefits of integrating quantum sensors into IoT networks are compelling, there are several technical and logistical challenges that must be addressed. One of the primary issues is the complexity and fragility of quantum technologies. Many quantum sensors require specific environmental conditions, such as low temperatures or isolation from electromagnetic interference, to function optimally. These constraints can limit their deployment in conventional IoT environments, which are often exposed to variable and uncontrolled conditions. Interoperability between quantum sensors and existing IoT infrastructure also presents a significant hurdle. Most IoT platforms are built around digital communication protocols and classical data processing architectures.

Quantum sensors, by contrast, may produce analog or quantum-encoded outputs that require specialized interfaces or quantum-aware algorithms for effective integration. Bridging this gap necessitates the development of new communication standards, middleware, and software tools that can translate and process quantum sensor data in an IoT context. Another challenge is power consumption and miniaturization. IoT devices are often battery-operated and require components that are energy-efficient and compact. Many current quantum sensors are relatively large and power-hungry, limiting their practicality for mobile or embedded applications. Ongoing research in quantum engineering and materials science is focused on developing portable and low-power quantum sensors suitable for widespread IoT deployment.

10.5.4 Security and Data Integrity in Quantum-Enhanced IoT

One of the key advantages of quantum technology integration in IoT networks lies in the realm of security. Quantum sensors, when coupled with quantum communication techniques such as quantum key distribution (QKD), can significantly bolster the cybersecurity of IoT systems. QKD ensures that any attempt to intercept or tamper with data can be detected, as it would alter the quantum state of the information being transmitted. This level of security is particularly important for critical infrastructure, financial systems, healthcare networks, and defense operations where data integrity is non-negotiable. Moreover, the use of quantum random number generators in IoT devices can enhance cryptographic protocols by providing truly random keys, thereby improving encryption strength. This ensures secure data transmission and storage, reducing vulnerabilities in IoT ecosystems that are often susceptible to cyberattacks due to weak encryption and poor access control.

10.5.5 Applications in Key Sectors

In the healthcare sector, the fusion of quantum sensors and IoT can revolutionize patient monitoring and diagnostics. Quantum biosensors, capable of detecting biomarkers at extremely low concentrations, can be used for early disease detection. When integrated into wearable IoT devices, these sensors can continuously monitor patients' vital signs and biochemical markers, transmitting data to healthcare providers in real time. This supports proactive healthcare and personalized treatment, especially for chronic disease management. In environmental monitoring, quantum sensors offer ultra-sensitive detection of pollutants, greenhouse gases, and radiation. Connected to IoT networks, these sensors can provide comprehensive and real-time environmental data, aiding in climate research, pollution control, and disaster preparedness. In smart city applications, quantum sensors can enhance traffic management, infrastructure integrity assessments, and energy grid optimization, contributing to safer and more sustainable urban environments. Industrial automation also benefits significantly from this integration. In manufacturing, quantum-enhanced IoT systems can detect minute changes in material stress, machine vibrations, or product deviations, allowing for high-precision quality control. These capabilities minimize downtime, reduce waste, and improve operational efficiency, key goals in Industry 4.0 frameworks.

CHAPTER 11

Quantum Sensing and the Future of Technology

11.1 THE ROLE OF QUANTUM SENSORS IN THE QUANTUM INTERNET

The evolution of modern communication networks is on the cusp of a transformative shift with the emergence of the quantum internet. Unlike traditional networks that rely on classical bits to transmit data, the quantum internet uses quantum bits, or qubits, to enable ultra-secure, high-speed communication based on the principles of quantum mechanics. A critical component in the realization of this futuristic network is the development and integration of quantum sensors. These sensors play a vital role not only in supporting the infrastructure and function of the quantum internet but also in enhancing its capabilities in terms of precision, security, and reliability. Quantum sensors are advanced measurement devices that exploit quantum mechanical effects to detect and measure physical quantities with unprecedented accuracy. These sensors can measure magnetic fields, temperature, pressure, gravitational forces, and time with sensitivity that surpasses classical technologies. The operation of quantum sensors is based on the quantum behavior of atoms, photons, or electrons. For example, atomic clocks use transitions between quantum energy levels to keep time with extraordinary precision, while superconducting quantum interference devices (SQUIDs) use quantum tunneling to measure minute magnetic fields. In the context of the quantum internet, these sensors can be used for timing synchronization, environmental monitoring, and even quantum state verification, all of which are crucial for the robust functioning of quantum networks.

11.1.1 Quantum Internet: Concept and Architecture

The quantum internet is envisioned as a network that enables the transmission of qubits across long distances without degradation of their quantum properties. This involves the use of quantum entanglement, teleportation, and quantum repeaters to distribute entangled particles and maintain quantum coherence over extended networks. Unlike classical internet, where data can be copied and resent, quantum information cannot be cloned due to the no-cloning theorem. As a result, maintaining the integrity of quantum states throughout transmission becomes a core challenge. The architecture of the quantum internet includes quantum nodes (which can be quantum computers, sensors, or memory units), quantum channels (often realized through fiber optics or free-space optics), and quantum repeaters that extend communication ranges by correcting errors and amplifying signals without disturbing the quantum states.

11.1.2 Role of Quantum Sensors in Timing and Synchronization

Precise timing is critical for the synchronization of quantum operations across distant nodes in the quantum internet. Quantum sensors, particularly atomic clocks, play a pivotal role in providing the ultra-accurate timing necessary for entanglement distribution and quantum teleportation protocols.

These clocks use the vibration frequency of atoms, such as cesium or rubidium, as a timekeeping standard and can maintain synchronization to within nanoseconds over long periods. This capability is vital for coordinating quantum key distribution (QKD), entanglement swapping, and error correction procedures, which require precise timing to ensure successful implementation. Without such timing fidelity, quantum communication would be vulnerable to decoherence and data loss.

11.1.3 Environmental Monitoring and Noise Reduction

Quantum sensors can also be employed to monitor and mitigate environmental factors that disrupt quantum communication. Quantum networks are highly sensitive to noise and interference from thermal fluctuations, magnetic fields, and vibrations.

Quantum sensors can detect these disturbances with extreme sensitivity, enabling real-time environmental monitoring and feedback-based compensation. For instance, nitrogen-vacancy (NV) centers in diamonds are used as magnetometers that can detect variations in magnetic fields at the nanoscale. When integrated into quantum internet nodes, these sensors can provide detailed information about the local environment, allowing the system to adapt dynamically and protect quantum coherence. This active monitoring enhances the robustness and fidelity of quantum communication.

11.1.4 Quantum Sensors as Network Nodes

Beyond supporting roles, quantum sensors can also function as integral components of the quantum internet by acting as quantum nodes themselves. These sensor nodes can participate in quantum information processing, sharing entangled states with other nodes and contributing to distributed quantum computing or sensing tasks. In distributed quantum sensing, multiple quantum sensors are entangled across the network to form a collective measurement system. This approach allows for measurements that surpass the standard quantum limit, achieving greater sensitivity and spatial resolution. Such networks could be used for geophysical exploration, navigation, or monitoring large-scale phenomena like seismic activity or gravitational waves with unprecedented precision.

11.1.5 Security and Quantum Key Distribution

Security is a foundational motivation for developing the quantum internet, and quantum sensors play an important role in supporting secure communication protocols. One of the most well-known applications is quantum key distribution (QKD), which enables two parties to share a cryptographic key with guaranteed security based on the laws of quantum mechanics. Quantum sensors can be used to verify the integrity of entangled states and detect any eavesdropping attempts. Any disturbance in the quantum state caused by interception will be noticed immediately, as it changes the measurement outcomes in a statistically significant way. Additionally, quantum sensors can ensure the proper alignment and calibration of quantum optical systems, which is crucial for secure and efficient QKD implementation.

11.1.6 Quantum Sensors in Quantum Repeaters

Quantum repeaters are essential components of the quantum internet, enabling the extension of entangled links over long distances by correcting errors and refreshing entanglement without collapsing quantum states. Quantum sensors contribute to the optimization of quantum repeater operation by providing precise measurements of signal quality, environmental stability, and timing. For example, optical quantum sensors can measure photon arrival times and polarization states with high accuracy, facilitating entanglement purification and swapping processes. Furthermore, the integration of quantum sensors into repeater nodes can improve error detection and feedback mechanisms, ensuring more reliable and efficient quantum communication over large-scale networks.

11.2 QUANTUM SENSING IN FUNDAMENTAL RESEARCH AND DISCOVERY

Quantum sensing represents one of the most promising frontiers in modern science, offering an unparalleled capability to detect and measure physical quantities with extreme precision. Leveraging the unique properties of quantum mechanics such as superposition, entanglement, and quantum tunneling, quantum sensors are able to surpass the limitations of classical measurement systems. These advancements are not only revolutionary for applied sciences but also play a crucial role in fundamental research and the pursuit of knowledge at the most basic levels of nature. From probing the mysteries of the universe to testing the boundaries of physical laws, quantum sensing has become an indispensable tool for discovery-driven science.

11.2.1 The Quantum Advantage in Measurement

At the heart of quantum sensing lies the exploitation of quantum states that respond to external stimuli with extraordinary sensitivity. Unlike classical systems, quantum systems can exist in multiple states simultaneously and exhibit correlations that defy classical explanations.

These properties allow quantum sensors to detect changes in time, magnetic fields, gravitational forces, and other physical parameters with astonishing resolution. For instance, quantum-enhanced atomic clocks provide timekeeping accuracy at the level of a few billionths of a second over millions of years, enabling tests of relativity and the constancy of fundamental constants. Such precision is essential in areas of fundamental research where minute variations can signal profound new insights into the workings of nature.

11.2.2 Probing the Foundations of Physics

Quantum sensing has significantly advanced efforts to test the fundamental laws of physics, particularly those related to general relativity and quantum field theory. By utilizing highly sensitive atomic interferometers and optical lattice clocks, researchers have conducted experiments to observe how time is affected by gravity at small scales, confirming Einstein's predictions with increasing accuracy. Moreover, these devices allow scientists to explore potential deviations from known laws, which may indicate the presence of new physical phenomena or previously undetected forces.

For example, ultra-precise sensors can detect possible violations of Lorentz invariance, a principle asserting that the laws of physics are the same for all observers regardless of their relative motion. Any observed violation could challenge our understanding of space-time and open new avenues in theoretical physics.

11.2.3 Exploring the Quantum Nature of Gravity

One of the most profound questions in fundamental physics is how gravity fits within the quantum framework. While quantum mechanics and general relativity are both well-tested theories, their integration into a single theory of quantum gravity remains elusive. Quantum sensors are emerging as valuable tools in this quest. Gravitational sensors based on atom interferometry can detect minute fluctuations in gravitational fields, making them suitable for investigating whether gravity itself has a quantum character. Recent proposals suggest that entangled quantum particles might be able to sense gravitational effects that would be undetectable by classical means, offering experimental pathways to probe the quantum aspects of gravity. These investigations are crucial for building a consistent theory that unifies all fundamental forces of nature.

11.2.4 Dark Matter and the Search for New Particles

Quantum sensing also plays a pivotal role in the search for dark matter and other elusive particles that do not interact with electromagnetic forces. Dark matter is believed to constitute a large portion of the universe's mass, yet it remains invisible and undetectable using conventional instruments. Quantum sensors offer a new method for detecting these hidden particles through their subtle effects on atomic and subatomic systems. For example, precision magnetometers and atomic clocks can be used to look for transient or spatially varying signals that may arise from interactions with dark matter fields. Similarly, axion detection experiments employ quantum technologies to detect the tiny electromagnetic signatures these hypothetical particles may produce. These applications exemplify how quantum sensing can address some of the most fundamental unanswered questions in cosmology and particle physics.

11.2.5 Measuring the Constants of Nature

The precision offered by quantum sensors has enabled scientists to measure fundamental constants such as the fine-structure constant, the gravitational constant, and the Planck constant with unprecedented accuracy. These constants govern the strength of interactions and the behavior of matter and energy across the universe. Any variation in their values could indicate new physics beyond the Standard Model. Atomic and optical sensors, in particular, are used in long-term experiments to monitor whether these constants change over time or under different gravitational potentials. Discovering even a tiny change could lead to a radical rethinking of our understanding of the universe and its evolution.

11.2.6 Understanding Condensed Matter and Exotic States

In condensed matter physics, quantum sensing facilitates the exploration of exotic phases of matter and quantum critical phenomena. By using sensors based on nitrogen-vacancy (NV) centers in diamond or superconducting qubits, researchers can examine magnetic, electric, and thermal properties at the nanoscale. These sensors are capable of imaging quantum materials with high spatial and temporal resolution, allowing for the study of phenomena such as superconductivity, topological states, and many-body quantum entanglement. Understanding these systems at a fundamental level not only deepens our theoretical knowledge but also paves the way for new technologies based on quantum materials and their unique properties.

11.2.7 Precision Navigation and Tests of Inertia

Quantum inertial sensors, such as accelerometers and gyroscopes based on cold atom interferometry, offer a highly accurate means of measuring motion and orientation. These devices are being used in experiments to test the equivalence principle, a cornerstone of general relativity that posits the equality of inertial and gravitational mass. By comparing the free fall of different atomic species under gravity, scientists can test this principle with extraordinary precision. Any observed discrepancy could hint at new interactions or the presence of additional dimensions. Furthermore, these precision instruments are valuable in geophysical research, helping to map Earth's gravitational field and internal structure with high resolution, which contributes to our understanding of planetary dynamics and tectonics.

11.2.8 Quantum Metrology and Standards

The field of quantum metrology, which underpins all scientific measurement, is undergoing a transformation driven by quantum sensing. Redefining measurement units based on fundamental constants, rather than physical artifacts, ensures universal consistency and long-term stability. For instance, the kilogram is now defined using the Planck constant via a device called the Kibble balance, which incorporates quantum electrical standards. Quantum sensors such as Josephson junctions and quantum Hall devices form the backbone of this new system of standards, allowing measurements of voltage, resistance, and current that are directly linked to fundamental physical constants. This approach not only improves measurement accuracy but also reinforces the theoretical coherence of the physical sciences.

11.3 THE POTENTIAL IMPACT OF QUANTUM SENSING ON INDUSTRY 4.0

The emergence of Industry 4.0 marks a significant transformation in the way industrial systems operate, characterized by the integration of cyber-physical systems, the Internet of Things (IoT), artificial intelligence (AI), and advanced data analytics. This new industrial paradigm emphasizes automation, smart manufacturing, and real-time data-driven decision-making. As the fourth industrial revolution gains momentum, the role of advanced sensing technologies becomes increasingly central to its success. Among these, quantum sensing stands out as a revolutionary approach capable of dramatically enhancing precision, efficiency, and reliability across various sectors.

11.3.1 Enhancing Precision Manufacturing and Quality Control

One of the most immediate and impactful applications of quantum sensing in Industry 4.0 is in precision manufacturing. The demand for ultra-fine tolerances and defect-free production processes is higher than ever, particularly in industries such as aerospace, semiconductors, automotive, and medical devices. Quantum sensors, which are capable of detecting minute variations in magnetic, electric, and gravitational fields, can be deployed to monitor machinery, tools, and materials with remarkable sensitivity. These capabilities allow for real-time error detection and correction, minimizing defects and maximizing product consistency. For example, quantum magnetometers can detect microscopic structural imperfections in metallic components that would otherwise go unnoticed using conventional techniques. Such high-resolution monitoring improves quality assurance processes and reduces waste, thereby increasing productivity and reducing operational costs.

11.3.2 Transforming Predictive Maintenance and Asset Monitoring

Predictive maintenance is a cornerstone of smart manufacturing under Industry 4.0, enabling the anticipation of equipment failures before they occur. Traditional sensors often fall short in identifying subtle early-warning signs of mechanical degradation. Quantum sensors, however, are capable of capturing extremely faint signals such as changes in vibration, thermal patterns, or electromagnetic emissions that precede machinery breakdowns. By integrating these sensors into industrial equipment, organizations can build robust predictive maintenance systems that rely on accurate, real-time data. This minimizes unplanned downtimes, extends the lifespan of assets, and enhances operational reliability. As industrial environments grow increasingly interconnected, quantum sensors can be embedded within IoT networks, contributing to a highly intelligent and self-aware factory ecosystem.

11.3.3 Revolutionizing Industrial Metrology and Standards

Precision measurement is a foundational element of manufacturing and process control. Quantum sensing has the potential to redefine industrial metrology by providing absolute measurements based on fundamental physical constants. Devices such as quantum gravimeters, atomic clocks, and interferometers offer reference standards that do not degrade over time, unlike traditional mechanical instruments. These sensors enable the precise calibration of tools, accurate alignment of machinery, and verification of material properties. In sectors such as nanotechnology and semiconductor manufacturing, where feature sizes continue to shrink, the role of quantum metrology becomes even more crucial. Accurate measurement at the nanoscale ensures adherence to specifications and supports the development of next-generation technologies.

11.3.4 Improving Logistics, Navigation, and Supply Chain Management

Efficient logistics and supply chain operations are essential for the competitiveness of modern industry. Quantum sensors, particularly those related to quantum inertial navigation and gravimetry, can offer significant improvements in positioning and tracking, especially in environments where GPS signals are unavailable or unreliable. Submarine navigation, underground mining, and warehouse automation are examples where quantum gyroscopes and accelerometers can maintain high-precision tracking. In logistics, quantum-enhanced sensors can be used to monitor environmental conditions such as temperature, pressure, and vibration during the transport of sensitive goods. These capabilities ensure the integrity of pharmaceutical products, perishable food, or high-value electronics throughout the supply chain, reducing spoilage and ensuring compliance with regulatory standards.

11.3.5 Enabling Advanced Materials Discovery and Testing

Materials science is at the core of many industrial innovations, from developing lightweight composites to high-strength alloys and nanostructured surfaces. Quantum sensing enables researchers and engineers to probe the intrinsic properties of materials at atomic and subatomic levels. Sensors based on nitrogen-vacancy (NV) centers in diamond, for instance, can detect magnetic and electric fields with nanoscale resolution. This allows for the exploration of material behaviors under different stress, thermal, and electromagnetic conditions, which is vital for designing robust industrial materials. These sensors can also be employed in nondestructive testing, enabling the inspection of structural integrity without compromising the material or component. This leads to safer and more efficient production environments.

11.3.6 Strengthening Environmental Monitoring and Industrial Safety

With increasing environmental regulations and sustainability goals, industries are under pressure to monitor emissions, pollutants, and hazardous conditions accurately. Quantum sensors can significantly enhance environmental monitoring by detecting trace gases, airborne particulates, and radiation with superior sensitivity. This enables industries to maintain compliance with environmental standards and respond promptly to anomalies. In terms of industrial safety, quantum sensors can be used to detect toxic leaks, structural stress, and fire hazards before they pose serious risks. These early warning capabilities not only protect human workers but also prevent costly damage to infrastructure and equipment. Moreover, quantum-enhanced imaging techniques could be employed in surveillance and safety inspections within complex industrial facilities.

11.3.7 Optimizing Energy Production and Grid Management

Energy production, transmission, and storage represent another domain where quantum sensing could bring transformative change. In the context of Industry 4.0, the smart grid concept relies on precise monitoring of voltage, current, and phase across vast electrical networks. Quantum sensors, especially those designed for electric field sensing and phase measurement, could improve the accuracy of monitoring systems, leading to better fault detection and dynamic load balancing. In renewable energy systems, such as wind and solar farms, quantum sensors can optimize the positioning and performance of generation units by providing real-time data on environmental and operational conditions. In oil and gas exploration, quantum gravimeters and magnetometers can enhance subsurface mapping, thereby improving resource extraction efficiency while minimizing environmental impact.

11.3.8 Accelerating Innovation in Automation and Robotics

Automation and robotics lie at the heart of Industry 4.0, where smart systems are designed to work autonomously and adaptively. The performance of these systems depends heavily on the quality of sensor input they receive. Quantum sensors can dramatically improve the perception capabilities of robots and automated systems. Enhanced spatial awareness, precise motion tracking, and sensitive environmental feedback allow robots to perform delicate tasks with greater accuracy and responsiveness. For example, quantum-enhanced imaging can enable robots to navigate through visually complex or low-light environments with ease. This is particularly useful in sectors such as precision agriculture, surgical robotics, and automated assembly lines.

11.3.9 Facilitating Data-Driven Decision-Making and AI Integration

The integration of quantum sensors into industrial systems also benefits data-driven decision-making, a fundamental aspect of Industry 4.0. The high-fidelity data collected by quantum sensors improves the input quality for machine learning algorithms and predictive models. This results in more accurate forecasts, better anomaly detection, and optimized process control. The synergy between quantum sensing and artificial intelligence enhances the capabilities of digital twins, enabling more precise simulations of manufacturing processes, supply chains, or asset behavior. This integrated approach leads to smarter, more agile, and resilient industrial operations.

11.4 CONVERGENCE OF QUANTUM SENSING WITH OTHER EMERGING TECHNOLOGIES (E.G., AI, 5G)

The rapid advancement of emerging technologies in the 21st century is reshaping how humans interact with the physical and digital worlds. Among these, quantum sensing has garnered significant attention for its ability to detect extremely subtle variations in physical quantities such as magnetic fields, gravitational forces, temperature, and pressure. By exploiting the fundamental principles of quantum mechanics—namely superposition, entanglement, and quantum coherence—quantum sensors promise unmatched sensitivity and precision. However, the true transformative potential of quantum sensing becomes evident when it is integrated with other cutting-edge technologies such as artificial intelligence (AI), 5G networks, edge computing, and the Internet of Things (IoT). This convergence opens up possibilities for smarter, faster, and more efficient systems across diverse domains, including healthcare, manufacturing, environmental monitoring, and autonomous systems.

11.4.1 Quantum Sensing and Artificial Intelligence: A Symbiotic Relationship

Artificial Intelligence has become indispensable in managing and interpreting vast amounts of data generated across industries. When coupled with quantum sensing, AI can enhance the extraction of actionable insights from the highly sensitive and complex datasets produced by quantum devices. Quantum sensors often capture data that is not only incredibly precise but also subject to quantum-level noise and variability. AI algorithms, particularly those based on machine learning and deep learning, are adept at pattern recognition and anomaly detection in such challenging environments. By applying AI to quantum sensor data, systems can autonomously learn from real-time inputs, identify trends, and optimize performance. This integration is especially relevant in medical diagnostics, where quantum sensors can detect minute changes in biological signals such as brain activity or heart rhythms. AI can process this information to predict health risks or monitor the progression of diseases more accurately. Similarly, in environmental science, AI-enhanced quantum sensors can detect trace pollutants or radiation levels with greater precision, helping to forecast ecological threats or mitigate natural disasters. The mutual reinforcement of AI and quantum sensing lays the groundwork for intelligent systems that are capable of adaptive learning, real-time decision-making, and autonomous operation.

11.4.2 Leveraging 5G for High-Speed Quantum Sensor Networks

Fifth-generation (5G) wireless communication technology provides high-speed data transfer, ultra-low latency, and massive connectivity, which are critical features for enabling large-scale deployment of quantum sensors in real-world environments. When integrated with quantum sensing, 5G acts as the communication backbone that connects a distributed network of sensors, allowing for the seamless transmission of high-resolution data to centralized or cloud-based analytics platforms. This synergy is particularly vital in time-sensitive applications such as autonomous vehicles, smart grids, and disaster response systems, where rapid data relay and processing are essential. In the context of smart cities, for example, a network of quantum sensors distributed across infrastructure could continuously monitor structural integrity, environmental conditions, and traffic flows. The 5G network would then ensure that this data is transmitted instantly to control centers for analysis and intervention. The convergence of quantum sensing and 5G thereby enables a new class of intelligent, connected, and responsive urban ecosystems. Moreover, the low-latency communication facilitated by 5G is essential for real-time feedback loops in quantum-enhanced robotics and industrial automation, where immediate action based on sensor data can prevent system failures or improve process efficiency.

11.4.3 Edge Computing and Decentralized Quantum Sensing Applications

The proliferation of edge computing—where data processing occurs near the source rather than in centralized servers—complements the capabilities of quantum sensors in decentralized environments. Many quantum sensing applications, such as autonomous navigation, battlefield surveillance, or wearable health monitoring, require data to be processed locally due to bandwidth, latency, or security constraints. By deploying AI models and analytics capabilities directly at the edge, near the quantum sensors, systems can make rapid decisions without relying on distant cloud infrastructure. For example, an autonomous drone equipped with quantum sensors can detect magnetic anomalies or gravitational variations while navigating a complex terrain. Edge computing enables the drone to process this data in real time, adjusting its path autonomously. Similarly, in healthcare, wearable quantum devices can monitor neurological or cardiac signals continuously and trigger alerts or interventions without waiting for cloud-based analysis. This localized intelligence enhances the practicality and responsiveness of quantum sensing technologies in the field, particularly in situations where real-time action can be life-saving or mission-critical.

11.4.4 Quantum Sensing and the Internet of Things: Towards Hyper-Connected Intelligence

The Internet of Things refers to the vast network of interconnected devices and sensors that communicate and interact autonomously. Integrating quantum sensors into IoT ecosystems has the potential to elevate the accuracy, reliability, and functionality of such networks. Traditional sensors in IoT systems are often limited in their sensitivity and subject to drift or interference. Quantum sensors, by contrast, offer extremely precise and stable measurements, which can improve the quality and trustworthiness of data across IoT applications. This convergence is highly impactful in industrial IoT, where precision monitoring of machinery, materials, and environments is crucial. For instance, quantum-enhanced sensors can detect the early stages of equipment wear or failure, feeding this information into an IoT network for real-time maintenance scheduling. In agricultural settings, quantum sensors can monitor soil moisture, nutrient content, and environmental conditions at an atomic scale, enabling precision farming practices that enhance crop yield and resource efficiency. The seamless integration of quantum sensing into IoT not only enhances the network's capability but also facilitates predictive analytics and intelligent automation across sectors.

11.4.5 Synergizing Quantum Sensing with Blockchain for Data Integrity

As data security and integrity become increasingly critical, especially in sensor-driven ecosystems, blockchain technology offers a decentralized solution for tamper-proof data storage and transmission. When quantum sensors are used to generate critical or sensitive data—such as in defense, aerospace, or medical applications—the combination with blockchain ensures that the data remains immutable and verifiable. Each data point collected by a quantum sensor can be recorded in a secure, distributed ledger, making it traceable and immune to unauthorized alteration. In smart supply chains, this integration can track the condition of high-value or sensitive goods from production to delivery. Quantum sensors can provide real-time information about temperature, humidity, and physical shocks, while blockchain ensures that this data is securely stored and accessible to authorized stakeholders only. This synergy enhances transparency, accountability, and trust in complex logistical operations, and can be extended to areas such as food safety, pharmaceutical logistics, and critical asset tracking.

11.5 THE ROADMAP TO COMMERCIALIZATION OF QUANTUM SENSORS

Quantum sensors are emerging as a transformative technology with the potential to revolutionize a wide array of industries, from healthcare and defense to energy, navigation, and environmental monitoring. These devices leverage the unique properties of quantum mechanics—such as superposition, entanglement, and quantum coherence—to achieve levels of sensitivity, accuracy, and precision far beyond those of classical sensors. Despite their groundbreaking capabilities, the path from laboratory innovation to market-ready products is complex and multifaceted. The roadmap to the commercialization of quantum sensors involves a coordinated effort across scientific research, engineering development, manufacturing scalability, regulatory compliance, and market adoption.

11.5.1 From Fundamental Research to Prototype Development

The foundation of quantum sensing lies in rigorous fundamental research. Academic institutions and government-funded laboratories have played a crucial role in pioneering the theoretical frameworks and experimental techniques that enable quantum sensing. These initial breakthroughs often involve controlled experiments in highly specialized laboratory environments. At this stage, researchers focus on isolating quantum effects and demonstrating proof-of-concept measurements using systems such as nitrogen-vacancy centers in diamonds, cold atoms, superconducting circuits, or trapped ions. While these demonstrations prove the potential of quantum sensors, they are typically too fragile, complex, or costly for immediate commercial application. The next step on the commercialization roadmap is transitioning from scientific discovery to functional prototypes. This involves engineering efforts to stabilize quantum systems, minimize environmental noise, and develop integrated control electronics and user interfaces. Prototypes are often customized for specific use cases, such as measuring gravitational anomalies for geophysical exploration or detecting subtle magnetic fields for medical diagnostics. These efforts require close collaboration between physicists, engineers, software developers, and systems designers to ensure that prototypes are both scientifically robust and practically operable in semi-controlled environments.

11.5.2 Technology Readiness and Scalability Challenges

Once a working prototype has been developed, attention shifts to increasing the technology readiness level (TRL). Moving from a laboratory-scale prototype to a commercial product involves overcoming significant challenges related to miniaturization, robustness, and scalability. Quantum sensors must be adapted to operate in real-world environments, which are far more unpredictable and noisy than laboratory settings. Engineers must develop packaging and shielding solutions that protect delicate quantum states while maintaining sensor performance and portability. Another major hurdle in this phase is ensuring reproducibility and consistency across multiple units. Manufacturing quantum sensors at scale requires high-precision fabrication techniques, often involving advanced materials and nanofabrication tools. Quality control becomes paramount, as small deviations in component properties can lead to significant variations in sensor performance. Developing standardized production processes and leveraging microelectromechanical systems (MEMS) or complementary metal-oxide-semiconductor (CMOS) technologies can help make quantum sensors more manufacturable and cost-effective.

11.5.3 Securing Funding and Building Industry Partnerships

The development and commercialization of quantum sensors are capital-intensive endeavors. From research to prototype to market-ready product, significant financial investment is needed to sustain innovation and scale production. This has led to the growing involvement of venture capital firms, corporate R&D divisions, and public-private partnerships in the quantum technology ecosystem. Governments around the world, recognizing the strategic importance of quantum technologies, have launched national initiatives and funding programs to support commercialization efforts. In parallel, forming partnerships between startups, academic institutions, and established industry players is crucial for accelerating commercialization. Startups often possess the agility and innovative mindset to push technological boundaries, while larger firms provide the infrastructure, market access, and business acumen required to bring products to scale. These collaborations facilitate the sharing of resources, expertise, and risk, thereby smoothing the transition from laboratory to market. Technology incubators, accelerators, and innovation hubs also play a role in supporting early-stage quantum sensing companies through mentorship, networking, and pilot opportunities.

11.5.4 Navigating Regulatory, Ethical, and Standardization Issues

As quantum sensors approach commercial deployment, regulatory and ethical considerations come into sharper focus. Depending on their application, quantum sensors may need to comply with existing standards in healthcare, defense, telecommunications, or industrial sectors. For example, a quantum-enabled magnetic resonance imaging (MRI) device would need to undergo rigorous clinical testing and receive regulatory approval from health authorities. Similarly, quantum navigation systems intended for military use must meet security, durability, and interoperability standards. The emerging nature of quantum technologies presents a unique challenge: in many cases, regulatory frameworks do not yet exist or are still evolving. Developing new standards and guidelines specific to quantum sensing requires dialogue among stakeholders, including researchers, manufacturers, regulators, and policymakers. Ethical considerations, such as data privacy, environmental impact, and dual-use concerns, must also be addressed proactively. Establishing clear frameworks for responsible development and use will not only facilitate regulatory compliance but also build public trust and confidence in quantum sensing technologies.

11.5.5 Market Validation and Application Demonstrations

Commercial success for quantum sensors hinges on their ability to address real-world problems more effectively than existing solutions. Demonstrating this value proposition to potential customers and stakeholders is essential for market validation. Pilot programs and field trials allow companies to test their quantum sensing products in practical environments, gather user feedback, and refine product features based on operational needs. These demonstrations provide evidence of reliability, scalability, and return on investment, which are crucial for convincing industries to adopt new technologies. Certain markets are particularly ripe for early adoption of quantum sensors. The defense and aerospace sectors, for instance, are interested in quantum gravimeters and magnetometers for GPS-independent navigation and submarine detection. In healthcare, quantum sensors could enhance brain imaging or cardiac diagnostics. The oil and gas industry sees potential in quantum gravimetry for more accurate subsurface mapping. By targeting niche, high-value applications first, companies can build a revenue base and gather performance data, which can be leveraged for broader market expansion later.

CHAPTER 12

Best Practices for Developing Quantum Sensing Systems

12.1 SYSTEM DESIGN FOR QUANTUM SENSOR INTEGRATION

Quantum sensors represent a groundbreaking advancement in precision measurement, leveraging the unique properties of quantum mechanics to achieve unparalleled sensitivity and accuracy. These sensors are poised to transform numerous industries, including healthcare, environmental monitoring, defense, and telecommunications. However, to realize the full potential of quantum sensors, a well-structured system design is essential. Integrating quantum sensors into practical applications involves addressing a range of technical challenges, such as signal processing, noise reduction, sensor calibration, and interfacing with existing technologies. A carefully designed system must account for the complexities of quantum mechanics while ensuring that the sensors can operate reliably in real-world environments. Before delving into the integration process, it is crucial to understand the fundamental properties of quantum sensors.

These devices typically rely on quantum states—such as superposition and entanglement—to measure physical quantities with extreme precision. Unlike classical sensors, which may rely on mechanical or electronic systems, quantum sensors can achieve sensitivity far beyond the capabilities of traditional technologies. This includes measuring very weak magnetic fields, gravitational waves, and minute temperature changes. Quantum sensors often work by measuring tiny changes in the quantum state of a system, and these changes can provide highly accurate data about the environment in which the sensor operates.

For successful integration, it is important to understand the quantum phenomena that govern the sensor's operation. Quantum coherence, for instance, is essential for achieving the high precision typical of quantum sensors. However, this property also makes quantum sensors susceptible to decoherence from environmental factors like temperature fluctuations and electromagnetic interference. A key aspect of system design, therefore, is ensuring that the quantum sensor can operate in conditions where these external influences are minimized or compensated for, making it critical to design systems that preserve the delicate quantum states over extended periods.

12.1.1 Key Design Considerations for Quantum Sensor Systems

The integration of quantum sensors into practical systems requires careful consideration of several key factors. These include environmental stability, data acquisition, sensor calibration, signal processing, and system interfaces. Each factor plays a significant role in determining the sensor's overall performance and reliability.

Environmental Control and Stability

Quantum sensors are particularly sensitive to environmental conditions. Factors like temperature, magnetic fields, vibrations, and humidity can easily disrupt the delicate quantum states required for accurate measurements. A significant challenge in system design is ensuring that the sensor environment is stable and controlled. For instance, many quantum sensors require extremely low temperatures to maintain coherence, necessitating the use of cryogenic systems or specially designed temperature-regulated enclosures.

Additionally, sensors like atomic interferometers or magnetometers require shielding to protect against external magnetic fields and to reduce noise. The system design must integrate these environmental control elements seamlessly into the sensor's operation. This includes incorporating sophisticated thermal management solutions, vibration dampening, and electromagnetic shielding to protect the sensor from interference.

For example, in quantum magnetometry, it may be necessary to use superconducting materials that operate at very low temperatures, which adds a layer of complexity to the system design in terms of power consumption and cooling requirements.

Sensor Calibration and Precision

For quantum sensors to deliver the precision they promise, they must be carefully calibrated. Calibration ensures that the sensor's output accurately reflects the physical quantities being measured. However, quantum sensors are inherently more sensitive than classical sensors, meaning that even slight deviations from ideal conditions can lead to significant measurement errors.

A well-designed system must include mechanisms for continuous calibration, either through built-in reference standards or periodic recalibration processes. Additionally, calibration in quantum sensor systems is often more complex than in traditional systems. For example, the application of quantum feedback techniques or the use of auxiliary quantum systems may be necessary to correct errors induced by noise or other environmental factors. The system design should also account for potential drift in the sensor's measurements over time and incorporate methods for real-time correction.

Signal Processing and Noise Reduction

Quantum sensors generate data that must be processed and interpreted, often in real-time, to extract meaningful information. However, one of the challenges of quantum sensing is the inherent noise present in the measurement process. Quantum systems are sensitive not only to external environmental factors but also to intrinsic noise due to quantum fluctuations. For example, quantum noise, such as shot noise or phase noise, can obscure the signals being measured. Signal processing techniques are essential to filter out this noise and enhance the precision of the sensor's readings. In a well-designed quantum sensor system, advanced signal processing algorithms must be incorporated to extract useful information from noisy data. This may involve the use of techniques like error correction, quantum filtering, and post-processing to improve the signal-to-noise ratio. Additionally, the system must include hardware that can support the computational load associated with real-time signal processing, which may include powerful processors or specialized quantum computers for more complex tasks.

Integration with Existing Infrastructure

To be truly functional, quantum sensors must integrate smoothly with existing infrastructure, particularly in industries like telecommunications, defense, and healthcare, where legacy systems are already in place. This requires the system design to consider compatibility with current technology standards, communication protocols, and data formats. In practical terms, this might involve the integration of quantum sensors into larger sensor networks or data acquisition systems that interface with cloud platforms or edge computing devices.

Quantum sensors can be particularly valuable in fields like healthcare, where they can improve diagnostic tools and precision medicine. In such cases, it is important that quantum sensors interface with existing medical equipment, such as MRI machines or diagnostic software platforms. The system design should prioritize ease of integration, with standardized interfaces and communication protocols that allow quantum sensors to function alongside conventional technologies.

Power Consumption and Scalability

One of the major challenges in the integration of quantum sensors into real-world systems is power consumption. Many quantum sensors require complex support systems—such as cryogenic cooling or vacuum chambers—to maintain the necessary quantum conditions. These additional systems can be energy-intensive, making it essential to design quantum sensor systems with energy efficiency in mind. Furthermore, scalability is an important consideration. In applications like industrial monitoring or large-scale environmental sensing, large arrays of quantum sensors may need to operate simultaneously, often in remote or hard-to-reach locations. The system design must account for the scalability of the technology, ensuring that the infrastructure can support multiple sensors operating in parallel without compromising performance. This may include designing wireless communication systems, optimizing power consumption, and ensuring that the system can handle a large volume of data generated by multiple sensors.

12.1.2 The Role of Software and User Interfaces

A successful quantum sensor system goes beyond hardware—it also includes software that allows users to interact with the system and interpret the data. The software component of the system design is critical for ensuring that the data collected by quantum sensors is accessible, understandable, and actionable. This includes the development of user-friendly interfaces that display real-time sensor data, alert systems for abnormal readings, and advanced analytics tools that provide deeper insights. In many applications, users will need to access data from multiple sensors and interpret it in the context of broader systems. For example, in environmental monitoring, data from quantum sensors must be integrated with other sensor types, such as temperature or humidity sensors, to provide a complete picture of environmental conditions. This requires seamless integration between different types of sensors and sophisticated software tools for data fusion and visualization.

12.2 CALIBRATION AND VALIDATION OF QUANTUM SENSORS

Quantum sensors have emerged as a revolutionary technology, leveraging the principles of quantum mechanics to offer unprecedented precision in measurements of physical quantities such as temperature, magnetic fields, time, and acceleration. While the capabilities of quantum sensors promise groundbreaking applications across various fields like healthcare, telecommunications, and environmental monitoring, their full potential can only be realized if they are properly calibrated and validated. Calibration and validation are critical steps in ensuring the reliability, accuracy, and consistency of quantum sensors in real-world applications.

12.2.1 Understanding Calibration and Validation in the Context of Quantum Sensors

Calibration refers to the process of adjusting a sensor to provide accurate readings by comparing it to known standards or reference values. For quantum sensors, calibration ensures that the sensor's output correctly reflects the physical quantities it is measuring. This process often involves subjecting the sensor to a series of controlled measurements with known inputs and making adjustments to the sensor's internal parameters to align its readings with those inputs. Validation, on the other hand, involves verifying the performance of a sensor in real-world conditions. It is the process of assessing the sensor's accuracy, repeatability, and robustness by comparing its measurements against established benchmarks or ground truth data under actual operational conditions. While calibration is typically conducted in a controlled laboratory setting, validation is conducted in situ, under conditions that closely resemble the sensor's intended use. For quantum sensors to fulfill their potential, both calibration and validation must be performed meticulously. Any deviation from accurate measurements due to improper calibration or failure to account for environmental factors during validation could lead to unreliable data, undermining the sensor's effectiveness in practical applications.

12.2.2 The Calibration Process for Quantum Sensors

The calibration of quantum sensors requires a deep understanding of both the quantum principles behind the sensor's operation and the practical challenges involved in measuring physical quantities accurately. Unlike classical sensors, which operate based on well-understood physical laws, quantum sensors exploit quantum phenomena like superposition, entanglement, and quantum coherence, all of which can be susceptible to various sources of error. One of the key steps in the calibration of quantum sensors is the determination of the sensor's sensitivity to various types of physical quantities.

For example, a quantum magnetometer may need to be calibrated to accurately measure magnetic fields by subjecting it to known magnetic field strengths. Similarly, a quantum thermometer requires calibration against temperature references, such as fixed-point temperature standards, to ensure that it delivers accurate temperature readings. A significant challenge in the calibration of quantum sensors is minimizing the impact of environmental noise. Quantum sensors are highly sensitive to fluctuations in temperature, electromagnetic fields, and other external factors, which can cause errors in measurement. As a result, calibration is often conducted in controlled environments, such as cryogenic chambers or vacuum systems, to reduce external influences.

The process may also involve applying quantum error correction techniques, where quantum algorithms are used to compensate for certain types of errors or noise introduced during the calibration process. Another consideration in the calibration of quantum sensors is the relationship between the sensor's quantum state and the physical quantity being measured. In many cases, quantum sensors work by measuring changes in the quantum state, such as phase shifts or population differences in atomic or molecular systems. The calibration process must ensure that the sensor's quantum state accurately reflects the physical parameters it is designed to measure, which often requires precise mathematical modeling and extensive experimentation.

12.2.3 Challenges in the Calibration of Quantum Sensors

Calibration of quantum sensors is not without its challenges. One major challenge is the fragile nature of quantum systems. Quantum coherence, a key feature of quantum sensors, is sensitive to even small disturbances from the environment. As a result, any deviation from the ideal operating conditions during calibration can cause significant errors in measurements. For example, even slight variations in temperature or electromagnetic fields can disrupt the quantum states of particles, leading to calibration inaccuracies. Moreover, the need for highly controlled environments—such as cryogenic temperatures or low-pressure conditions—can make calibration difficult and expensive. Many quantum sensors, such as those based on atomic clocks or superconducting qubits, require ultra-low temperatures to maintain the quantum states needed for accurate measurements. These specialized environments can increase the complexity and cost of calibration, limiting accessibility for widespread adoption.

12.2.4 Validation of Quantum Sensors in Real-World Conditions

While calibration ensures that a quantum sensor provides accurate measurements under controlled conditions, validation is necessary to determine how the sensor performs in actual applications. Quantum sensors are often deployed in dynamic and unpredictable environments, where variables such as temperature fluctuations, vibrations, and electromagnetic interference can affect sensor performance. Validation ensures that the sensor can operate reliably in these real-world conditions. The process of validation involves testing the sensor in operational environments, such as laboratories, manufacturing facilities, or even remote locations, to assess its accuracy, precision, and robustness. Validation also entails evaluating the sensor's ability to maintain performance over time, ensuring that it provides consistent measurements despite potential sources of degradation like noise or drift. In quantum sensor validation, it is important to consider factors like the sensor's response time, sensitivity to environmental disturbances, and repeatability of measurements. Additionally, validation often involves comparing the quantum sensor's readings with those from conventional sensors or other established measurement techniques. In some cases, validation may require a series of real-world field tests to ensure that the quantum sensor is suited to its intended application. One of the challenges in the validation process is ensuring that the quantum sensor can continue to operate reliably despite potential disruptions to its quantum coherence. Since quantum sensors depend on maintaining coherence among quantum states, any loss of coherence—due to thermal fluctuations, electromagnetic noise, or other environmental factors—can cause the sensor's performance to degrade. As such, validating a quantum sensor's resilience to such disturbances is crucial for confirming its practical utility.

12.2.5 Technological Solutions for Calibration and Validation

As quantum sensors continue to evolve, researchers are developing a variety of technological solutions to address the challenges of calibration and validation. One promising approach is the use of feedback control systems that can adjust a sensor's parameters in real-time, helping to compensate for environmental noise and fluctuations. These systems allow for continuous monitoring and adjustment of the sensor's calibration during operation, improving long-term accuracy and stability. Another solution involves the development of quantum error correction protocols that help mitigate the effects of noise and decoherence. These protocols enable the quantum sensor to recover from errors, ensuring that the measurements remain reliable even in challenging conditions. In combination with advanced signal processing techniques, these error correction methods improve the overall performance of quantum sensors in both the calibration and validation phases. Furthermore, advances in simulation and modeling techniques are playing a crucial role in improving the calibration and validation of quantum sensors. By simulating quantum sensor systems under various operational conditions, researchers can better understand how environmental factors affect sensor performance and design more effective calibration and validation procedures. These simulations also provide valuable insights into the behavior of quantum sensors, which can guide the development of new technologies and methods for ensuring accurate measurements.

12.3 OVERCOMING TECHNOLOGICAL BARRIERS IN QUANTUM SENSING DEVELOPMENT

Quantum sensing, an emerging frontier in the application of quantum mechanics, has the potential to redefine precision in measurement across a wide range of fields, including healthcare, navigation, defense, environmental monitoring, and materials science. By exploiting quantum phenomena such as superposition, entanglement, and quantum tunneling, quantum sensors promise unparalleled sensitivity and accuracy. However, the journey from laboratory experiments to real-world, commercially viable devices is impeded by several technological barriers. These obstacles range from material limitations and environmental vulnerabilities to scalability and integration challenges. Overcoming these barriers is critical to unlocking the full potential of quantum sensing in both scientific and industrial landscapes.

12.3.1 Material Challenges and Innovations

One of the foremost barriers in quantum sensor development lies in the availability and manipulation of suitable quantum materials. Many quantum sensors rely on specific atomic or subatomic properties that require rare or highly engineered materials. For example, nitrogen-vacancy (NV) centers in diamond are widely used for magnetic field sensing due to their stable spin properties at room temperature. However, the fabrication of high-quality diamond with consistent NV center characteristics remains expensive and technically complex. Similarly, superconducting materials used in quantum interference devices often require ultra-low temperatures, typically achieved using liquid helium, which poses challenges in terms of cost and portability. To overcome material-related limitations, researchers are investing in the development of alternative quantum materials that offer greater stability, lower cost, and ease of fabrication. Materials like silicon carbide, rare-earth ion crystals, and ultracold atomic gases are being explored for their quantum properties and compatibility with existing semiconductor technologies. Advances in nanofabrication and synthetic chemistry are also enabling more precise control over quantum defects, improving consistency and performance in sensor output. As material science progresses, these innovations are expected to reduce manufacturing costs and enhance the scalability of quantum sensors.

12.3.2 Environmental Sensitivity and Isolation

Quantum systems are inherently sensitive to external disturbances, including temperature fluctuations, electromagnetic interference, and mechanical vibrations. While this sensitivity enables highly precise measurements, it also makes quantum sensors vulnerable to environmental noise that can degrade performance or completely destabilize the quantum states on which the sensors rely. Maintaining coherence—the preservation of quantum superposition—is particularly challenging outside of controlled laboratory conditions. Addressing this issue requires the implementation of sophisticated isolation and shielding techniques. Cryogenic cooling systems, magnetic shielding, and vibration damping are commonly used to minimize environmental interference. However, these methods can be bulky, energy-intensive, and impractical for field applications. Researchers are therefore working to enhance the robustness of quantum sensors by developing noise-tolerant designs and incorporating quantum error correction protocols. These approaches enable the sensor to identify and correct for certain types of interference, thereby maintaining performance in more dynamic or uncontrolled environments. Additionally, miniaturized isolation technologies and robust packaging solutions are being engineered to make sensors more portable and rugged without compromising sensitivity.

12.3.3 Scalability and Manufacturability

Transitioning from experimental prototypes to scalable commercial products is one of the most significant hurdles in the development of quantum sensing technologies. Many quantum sensors are currently hand-built in research laboratories, involving complex alignment and calibration procedures that are not conducive to mass production. This lack of standardization hinders efforts to reduce costs, improve reliability, and ensure consistent performance across multiple units. To address scalability challenges, industry stakeholders are focusing on the development of standardized fabrication techniques and modular architectures. Integrated photonics, for instance, offers a promising route to scalability by enabling the integration of quantum components on a single chip using semiconductor manufacturing methods. Microelectromechanical systems (MEMS) are also being explored for their potential to host quantum sensing elements in compact, scalable formats. Collaborative initiatives between academia, industry, and government agencies are fostering the creation of fabrication standards and quality control frameworks, which are essential for building a reliable supply chain for quantum sensor components.

12.3.4 Integration with Classical Systems

For quantum sensors to be adopted in practical applications, they must interface seamlessly with classical data acquisition, processing, and control systems. This integration poses significant technical challenges due to differences in data formats, signal levels, and operating conditions. Quantum sensors often produce weak analog signals that require sophisticated amplification and filtering before they can be interpreted by conventional electronics. Additionally, the control systems needed to initialize, manipulate, and read out quantum states must operate with high precision and timing accuracy, often at cryogenic temperatures or in ultra-high vacuum conditions. Efforts to overcome these challenges involve the co-development of hybrid quantum-classical systems, where classical electronics are adapted or redesigned to support quantum sensor operations. Advanced analog-to-digital converters, low-noise amplifiers, and high-speed data links are being developed to meet the unique requirements of quantum sensing applications. Moreover, software frameworks that bridge quantum hardware with classical computing infrastructure are being designed to facilitate real-time data analysis, calibration, and control. These developments are critical to enabling end-users in sectors such as healthcare, defense, and industry to effectively utilize quantum sensors without requiring in-depth expertise in quantum physics.

12.3.5 Operational Stability and Reproducibility

Another major hurdle in the deployment of quantum sensors is ensuring their long-term operational stability and reproducibility across different environments and users. While laboratory demonstrations can achieve high levels of sensitivity under ideal conditions, real-world applications demand consistent performance over extended periods and under variable operating conditions. Sensor drift, aging of materials, and calibration dependencies can all undermine long-term reliability. To tackle these issues, researchers are designing quantum sensors with self-monitoring and self-correcting capabilities. For instance, real-time feedback loops and adaptive algorithms can be used to continuously tune the sensor parameters and maintain optimal operating conditions. Built-in calibration routines that reference known standards can help maintain accuracy over time. Furthermore, automated quality assurance procedures during manufacturing are essential to ensure that each unit performs within specified tolerances. As sensors move into broader deployment, rigorous validation protocols and lifecycle testing will be vital to establishing trust in their long-term reliability.

12.3.6 Economic and Infrastructural Considerations

Beyond technical barriers, economic and infrastructural challenges also affect the development and adoption of quantum sensing technologies. High costs associated with research, materials, and fabrication limit access to quantum sensors and slow their commercialization. Moreover, a lack of infrastructure—such as testing facilities, quantum-ready manufacturing lines, and trained personnel—can impede progress.

To address these issues, governments and private enterprises are investing in quantum innovation hubs, pilot programs, and education initiatives aimed at building a skilled workforce and supportive ecosystem. National quantum strategies in countries like the United States, Germany, China, and India are directing funding toward infrastructure development and cross-sector collaboration. By fostering public-private partnerships and incentivizing industry participation, these efforts aim to create a more accessible and economically viable pathway for quantum sensor development and deployment.

12.3.7 The Path Forward: Research and Collaboration

Overcoming the technological barriers in quantum sensing development requires sustained interdisciplinary research and international collaboration. Physicists, engineers, materials scientists, computer scientists, and industry practitioners must work together to address the multifaceted challenges associated with these devices. Open-access research, shared standards, and collaborative platforms are essential to accelerating innovation and translating laboratory advances into real-world solutions. Emerging trends such as the convergence of quantum sensing with artificial intelligence, machine learning, and edge computing are expected to further expand the capabilities of quantum sensors and improve their resilience to operational challenges. These synergies can enable smarter data processing, autonomous calibration, and predictive maintenance, all of which contribute to more robust and user-friendly sensing systems.

12.4 CROSS-DISCIPLINARY COLLABORATION IN QUANTUM SENSING

Quantum sensing, a rapidly advancing field that leverages quantum mechanical phenomena such as superposition, entanglement, and quantum tunneling, offers extraordinary precision in measuring physical quantities such as time, magnetic fields, temperature, and gravitational forces. As the boundaries of classical sensing reach their limits, quantum sensing opens up new possibilities across numerous domains. However, the development and application of quantum sensors require more than the expertise of physicists alone.

The field thrives on robust cross-disciplinary collaboration, drawing from areas including physics, engineering, computer science, materials science, biomedical research, and industrial design. The intersection of these diverse disciplines is not merely a convenience but a necessity to fully realize the potential of quantum sensing in real-world scenarios.

12.4.1 The Role of Physicists and Quantum Theorists

At the heart of quantum sensing lies the foundational knowledge of quantum mechanics. Physicists, especially those specializing in quantum theory, are the architects of sensor design principles. They develop theoretical models to understand and predict how quantum phenomena can be exploited for sensing applications. The realization of entangled states, coherence time maintenance, and manipulation of qubits for sensing tasks are central to their work.

These theoretical insights serve as the blueprint from which practical sensors can be constructed. Moreover, physicists contribute significantly to interpreting the fundamental limitations and possibilities within which quantum sensors operate, thus guiding other disciplines in feasible directions.

12.4.2 Engineering Expertise in System Design and Integration

Once the theoretical framework of a quantum sensor is established, engineers step in to transform abstract concepts into physical devices. Electrical engineers, mechanical engineers, and systems engineers play a critical role in developing the hardware needed to support quantum sensing.

This includes cryogenic systems for maintaining low temperatures, vacuum chambers for isolating quantum states, and compact electronic circuits for signal processing and control. Optical engineers are also crucial when the quantum sensor involves photonic components, such as in atomic clocks or interferometric sensors. Through the integration of quantum sensing elements with classical systems, engineers ensure that these devices are not only functional but also scalable, robust, and adaptable for use outside laboratory settings.

12.4.3 Materials Science Contributions

Quantum sensors often rely on specific material properties to function effectively. For instance, nitrogen-vacancy centers in diamond require the careful engineering of diamond lattices, while atomic sensors demand highly pure and stable atomic environments. Materials scientists contribute by developing and characterizing these quantum-compatible materials.

They explore methods for doping, structuring, and stabilizing materials at the nanoscale to preserve quantum coherence and reduce noise. Their work enables the production of high-performance quantum sensing components that meet the demanding requirements of sensitivity, reproducibility, and durability. The collaboration between materials science and quantum physics has also led to the discovery of novel materials that are paving the way for next-generation quantum sensors.

12.4.4 Computer Science and Quantum Information Integration

Quantum sensors generate massive amounts of data, often in real time and with high precision. Computer scientists and specialists in quantum information processing are essential for developing the algorithms and computational architectures necessary to handle this data efficiently. Their expertise is particularly valuable in implementing noise reduction strategies, error correction techniques, and machine learning models that enhance the interpretation and reliability of quantum sensor outputs.

In addition, computer scientists play a vital role in simulating quantum systems, which helps to predict sensor behavior under various conditions and optimize their design before physical implementation. The fusion of quantum sensing with artificial intelligence is an emerging frontier that promises to make these sensors more adaptive and intelligent in real-world applications.

12.4.5 Biomedical Applications and Collaboration

In the realm of healthcare, quantum sensors offer transformative potential in areas such as brain imaging, diagnostics, and biomarker detection. Biomedical researchers and clinicians contribute by identifying medical challenges that quantum sensors can address more effectively than classical technologies. For instance, magnetoencephalography using quantum sensors can detect minute magnetic fields generated by neuronal activity with higher spatial resolution than existing techniques. Collaboration between biomedical researchers and quantum technologists ensures that sensors are developed with a deep understanding of biological systems and clinical requirements. This synergy facilitates the design of devices that are not only scientifically sound but also practical and safe for patient use.

12.4.6 Environmental and Earth Sciences Involvement

Quantum sensing also plays a vital role in environmental monitoring, geophysics, and climate science. Sensors that can detect gravitational anomalies or minute changes in magnetic fields have applications in detecting underground water reserves, monitoring tectonic activity, and mapping environmental changes. Scientists in these domains bring crucial knowledge about the types of measurements needed, the environments in which sensors will operate, and the practical challenges associated with deploying technology in the field. Their collaboration with quantum sensor developers ensures that the devices are tailored to specific scientific objectives and can withstand the harsh conditions of real-world deployments.

12.4.7 Industrial Design and User Experience

As quantum sensing technologies move closer to commercialization, collaboration with industrial designers becomes increasingly important. These professionals focus on making quantum sensors user-friendly, aesthetically appealing, and suitable for integration into larger systems or consumer products. Their input ensures that sensors can be manufactured at scale, maintained with ease, and operated by non-expert users. Industrial design also involves considerations of ergonomics, portability, and durability, especially for applications in healthcare, aerospace, and defense. By bridging the gap between sophisticated quantum technology and practical human use, industrial designers help to facilitate widespread adoption and market success.

Glossary of Key Terms

Atom Interferometer: A sensor that uses interference patterns of atoms to measure gravitational and inertial forces.

Atomic Clock: A quantum sensor that uses the vibration frequency of atoms to keep extremely accurate time.

Atomic Vapor Cell: A container holding vaporized atoms, used in magnetometers and atomic clocks.

Bose-Einstein Condensate (BEC): A state of matter at near absolute zero, used in precision measurements in quantum sensing.

Coherence Time: The duration over which a quantum state maintains its coherence, crucial for sensor performance.

Cold Atoms: Atoms cooled to near absolute zero to reduce thermal noise, used in high-precision sensors.

Entanglement: A quantum state where particles remain correlated regardless of the distance separating them, used for enhancing sensor sensitivity.

Gravimetry: The measurement of gravitational fields, which can be enhanced using quantum sensors.

Heisenberg Limit: The fundamental quantum limit of measurement accuracy, better than the standard quantum limit.

Hybrid Quantum Systems: Devices that combine different quantum platforms (e.g., spin systems with photonics) for enhanced sensing capabilities.

Hyperfine Splitting: A small shift in atomic energy levels due to nuclear interactions, used in precise measurements.

Interferometry: A measurement technique using wave interference, enhanced using quantum particles.

Laser Cooling: A technique to cool atoms using laser light, enabling their use in quantum sensing.

Magnetoencephalography (MEG): A non-invasive brain imaging technique that can use quantum magnetometers.

NV Center: A nitrogen-vacancy center in diamond, used as a highly sensitive quantum sensor for magnetic and electric fields.

Optical Lattice: A structure formed by intersecting laser beams that traps atoms for quantum sensing applications.

Photonic Sensor: A sensor that uses quantum properties of light (photons) for detection and measurement.

Quantum Communication: The transmission of quantum information, often interconnected with sensor networks.

Quantum Control: Techniques used to manipulate quantum systems accurately for sensing and computation.

Quantum Cryptography: Secure communication using quantum principles, occasionally integrated with sensing systems.

Quantum Decoherence: The loss of quantum coherence due to interaction with the environment, leading to classical behavior.

Quantum Dot: A nanoscale semiconductor particle with quantum properties, used in some sensing applications.

Quantum Enhancement: The use of quantum properties to achieve better performance than classical limits.

Quantum Field Sensor: A device that measures weak electromagnetic or gravitational fields using quantum phenomena.

Quantum Frequency Standard: A frequency reference derived from quantum systems, foundational for timekeeping.

Quantum Gravimeter: A sensor that measures gravitational acceleration using quantum interference.

Quantum Gyroscope: A rotation sensor based on quantum principles, offering improved accuracy over classical gyroscopes.

Quantum Imaging: An imaging technique that uses quantum correlations to enhance resolution and sensitivity.

Quantum Interference: The phenomenon where quantum wavefunctions combine, used to extract information with high precision.

Quantum Lithography: A technique using entangled photons to enhance spatial resolution, with applications in sensor fabrication.

Quantum Magnetometer: A device that detects magnetic fields with high sensitivity using quantum properties of atoms or spins.

Quantum Metrology: The science of measurement using quantum theory to improve precision.

Quantum Noise: The inherent fluctuations in quantum systems that can limit sensor precision.

Quantum Readout: The process of extracting information from a quantum system for sensing purposes.

Quantum Sensor: A device that uses quantum phenomena such as superposition and entanglement to achieve high precision in measuring physical quantities.

Quantum Simulation: Using quantum systems to simulate other physical systems, helping to model sensor behavior.

Quantum Thermometry: The use of quantum sensors to measure temperature with extremely high accuracy.

Quantum Tomography: A method of reconstructing the quantum state of a system, aiding in sensor diagnostics.

Quantum Tunneling: The ability of particles to pass through energy barriers, used in quantum magnetometers and sensors.

Qubit: The fundamental unit of quantum information, analogous to the classical bit but capable of being in superposition states.

Sensitivity: The ability of a sensor to detect small changes in a physical quantity.

Sensor Calibration: The process of adjusting a sensor to ensure accurate readings, essential in quantum sensing.

Shot Noise: A type of quantum noise arising from the discrete nature of particles, important in sensor design.

Single-Photon Detector: A sensor capable of detecting individual photons, important in quantum optics and sensing.

Spintronics: A field that exploits the spin property of electrons, relevant for developing quantum sensors.

Squeezed States: Special quantum states with reduced uncertainty in one variable, used to surpass measurement limits.

Standard Quantum Limit (SQL): The typical sensitivity limit for measurements not using entanglement.

Superposition: A quantum principle where a particle exists in multiple states simultaneously until measured.

Time-of-Flight Measurement: A sensing method measuring the time it takes particles to travel a distance, improved by quantum methods.

Vacuum Fluctuations: Quantum fluctuations in empty space, relevant for certain ultra-sensitive sensors.

Reference

Advances in Quantum Sensing Technologies Publisher: Wiley Publication Year: 2020

Coherent Control of Diamond Defects for Quantum Information Science and Quantum Sensing edited by Mikhail D. Lukin, Markus Greiner, and Amir Yacoby Publisher: Springer Publication Year: 2014

Handbook of Quantum Sensing Publisher: Springer Publication Year: 2016

Quantum Communication, Quantum Networks, and Quantum Sensing by Ivan Djordjevic Publisher: Academic Press Publication Date: July 14, 2022

Quantum Dot Sensors: Technology and Commercial Applications by John Callan Publisher: Jenny Stanford Publishing Publication Year: 2013

Quantum Information Science: The New Frontier in Quantum Computation, Secure Communication, and Sensing by Klaus Petritsch Publisher: Arcler Education Inc. Publication Year: 2018

Quantum Metrology and Sensing Publisher: CRC Press Publication Year: 2018

Quantum Sensing and AI: Unlocking the Next Evolution in Technology by Tom Batchelor Publisher: Palmetto Publishing Publication Date: December 3, 2024

Quantum Sensing and Imaging: A Beginner's Guide to Sensing and Imaging by V.T. Harikumar Publisher: Draft2Digital Publication Year: 2024

Quantum Sensing and Measurement Publisher: Oxford University Press Publication Year: 2007

Quantum Sensing and Metrology (Chapter in "Law and Policy for the Quantum Age") Publisher: Cambridge University Press Publication Year: 2022

Quantum Sensing and Nano Electronics and Photonics XIV edited by Manijeh Razeghi Publisher: SPIE Publication Year: 2018

Quantum Sensing and Nanophotonic Devices XII edited by M. Razeghi Publisher: SPIE Publication Year: 2015

Quantum Sensing at the Interface of Nanotechnology Integrated Microfluidics by Kamil Reza Khondakar and Ajeet Kumar Kaushik Publisher: Elsevier Publication Year: 2023

Quantum Sensing for Defense and Security Publisher: Artech House Publication Year: 2013

Quantum Sensing in Nanotechnology Publisher: Springer Publication Year: 2005

Quantum Sensing in Space Applications Publisher: NASA Publications Publication Year: 2011

Quantum Sensing Technologies and Applications Publisher: Elsevier Publication Year: 2015

Quantum Sensing: Advances and Challenges Publisher: Elsevier Publication Year: 2004

Quantum Sensing: An Introduction Publisher: McGraw-Hill Publication Year: 2002

Quantum Sensing: Devices and Systems Publisher: Oxford University Press Publication Year: 1999

Quantum Sensing: Fundamentals and Applications Publisher: Wiley Publication Year: 2000

Quantum Sensing: Materials and Devices Publisher: World Scientific Publication Year: 2010

Quantum Sensing: Principles and Applications Publisher: Oxford University Press Publication Year: 2021

Quantum Sensing: Principles and Techniques Publisher: Springer Publication Year: 2001

Quantum Sensing: Techniques and Applications Publisher: Taylor & Francis Publication Year: 2003

Quantum Sensing: Theory and Practice Publisher: Springer Publication Year: 2009

Quantum Sensor Networks Publisher: Springer Publication Year: 2012

Quantum Sensors and Metrology Publisher: PS Publishing Publication Date: July 18, 2024

Quantum Sensors in Biomedical Applications Publisher: Taylor & Francis Publication Year: 2014

Quantum Sensors: A Comprehensive Guide Publisher: CRC Press Publication Year: 2006

Quantum Sensors: An Overview Publisher: Springer Publication Year: 2019

Quantum Sensors: Design and Implementation Publisher: McGraw-Hill Publication Year: 2017

Quantum-Enhanced Sensing Based on Time Reversal of Entangling Interactions by Daniel Linnemann Publisher: Springer Publication Date: August 9, 2018

The Quantum Sensing Handbook Publisher: PS Publishing Publication Date: July 25, 2024

Towards Scalable Quantum Sensors: Interface Electronics for Quantum Sensors by Michal Kern et al. Publisher: Now Publishers Publication Date: December 18, 2024

Index

G

H

I

J

K

L

M

N

O

P

Q

R

S

T

U

V

W

X

Y

Z

www.ingramcontent.com/pod-product-compliance
Lightning Source LLC
LaVergne TN
LVHW071245210926
848486LV00008B/350